T0269326

The two volumes that comprise this work provide a comprehensive guide and source book on the marine use of composite materials.

The first volume, *Fundamental Aspects*, provides a rigorous development of theory. Areas covered include materials science, environmental aspects, production technology, structural analysis, finite-element methods, materials failure mechanisms and the role of standard test procedures. An appendix gives tables of the mechanical properties of common polymeric composites and laminates in marine use.

The second volume, *Practical Considerations*, examines how the theory can be used in the design and construction of marine structures, including ships, boats, offshore structures and other deep-ocean installations. Areas covered in this second volume include design, the role of adhesives, fabrication techniques, and operational aspects such as response to slam loads and fatigue performance. The final three chapters of the book cover regulatory aspects of design, quality and safety assessment, and management and organisation.

These volumes will provide an up-to-date introduction to this important and fast-growing area for students and researchers in naval architecture and maritime engineering. It will also be of value to practising engineers as a comprehensive reference book.

CAMBRIDGE OCEAN TECHNOLOGY SERIES 4

General Editors: I. Dyer, R. Eatock Taylor, J. N. Newman and
W. G. Price

COMPOSITE MATERIALS IN MARITIME STRUCTURES

Volume 1: Fundamental Aspects

Cambridge Ocean Technology Series

COMPOSITE MATERIALS IN MARITIME STRUCTURES

Volume 1: Fundamental Aspects

Edited by

R. A. Shenoi

and

J. F. Wellicome

Department of Ship Science, University of Southampton

West
E uropean
G raduate
E ducation
M arine
T echnology

CAMBRIDGE
UNIVERSITY PRESS

CAMBRIDGE UNIVERSITY PRESS
Cambridge, New York, Melbourne, Madrid, Cape Town, Singapore, São Paulo, Delhi

Cambridge University Press
The Edinburgh Building, Cambridge CB2 8RU, UK

Published in the United States of America by Cambridge University Press, New York

www.cambridge.org
Information on this title: www.cambridge.org/9780521451536

First published 1993
This digitally printed version 2008

A catalogue record for this publication is available from the British Library

ISBN 978-0-521-45153-6 hardback
ISBN 978-0-521-08993-7 paperback

CONTENTS

PREFACE

The use of Fibre Reinforced Plastics (FRP) in the marine field has been growing steadily since the early 1950s. Initially FRP was used for small craft such as lifeboats and pleasure craft. This has changed over the years to the point where structures having a mass of several hundred tonnes are regularly produced and used. Potential applications range from small components such as radar domes, masts and piping to larger scale structures such as ship hulls, ship superstructures, submersibles and offshore structure modules.

Alongside this growth in the number and size of FRP applications have come advances in materials technology, production methods and design procedures. It is now possible in many instances to produce structures which out-perform metal structures in terms of weight, strength and cost. To achieve this performance, good quality control procedures are vital. Equally, it is necessary for the designer and producer to have an all-round knowledge of FRP composite materials and related mechanics.

This book and its companion volume, which deals with Practical Considerations, are intended to provide a sound, theoretical base for the design and manufacture of major load-bearing structural members fabricated from FRP composites and to illustrate, through case studies, the particular features of the use of FRP in the marine field. The material has been grouped together to form two companion volumes that may also be useful independently of each other.

Volume 1 is titled "Composite Materials in Maritime Structures - Fundamental Aspects". This book contains the fundamental materials sciences, a discussion of failure mechanisms in FRP and the theoretical treatment of failure in design. There is a discussion of the applied mechanics of complex, orthotropic laminates and the basis of strength calculations for single skin and sandwich structures. Attention is also given to numerical techniques such as finite element analysis.

Volume 2 is titled "Composite Materials in Maritime Structures - Practical Considerations". This book is devoted to applications of FRP in the maritime field. It has sections on the design of craft operating in a displacement mode or with dynamic lift and support. Production considerations related to single skin and sandwich structures are discussed. Operational aspects related to material and structural failure are covered. The book closes with an examination of the impact of regulations on quality and of the complexities of design management.

The authors of various chapters in the two books are all professionally engaged in the field of FRP composites technology working in the marine industry, Classification Societies, research establishments and universities. They are known internationally for their contributions to the subject. Without the tremendous effort put in by the authors and their cooperation in meeting deadlines, these books would not have been possible. The editors wish to thank the authors for all their help and assistance.

The material in these two books was compiled for the 18th WEGEMT Graduate School held in the University of Southampton in March 1993. WEGEMT is a European network of universities in Marine Technology which exists to promote continuing education courses in this broad field, to encourage staff and student exchanges among the 28 (current) members and to foster common research interests.

Participants at such schools have generally been drawn from the ranks of practising, professional engineers in shipyards, boatyards, offshore industries, design consultancies and shipping companies. A large proportion have also been postgraduate students and staff from various academic and research establishments wishing to obtain an overview of a particular topic as a basis for research.

It is intended that these two books should be of interest to a similar spectrum of readers and that they could be used as a text for advanced undergraduate or postgraduate courses in FRP technology. Furthermore, they could be used as a basis for continuing education courses for young engineers in industry. Although aimed primarily at the marine field, the material in these books would also be relevant in other contexts where FRP is used for structural purposes.

The 18th WEGEMT School was organised with the help of an International Steering Committee whose members were:

Mr. O. Gullberg   Karlskronavaarvet AB, Sweden
Mr. A. Marchant   AMTEC, UK
Prof. M.K. Nygard Veritas Research & University of Oslo, Norway
Prof. H. Petershagen University of Hamburg, Germany

Dr. R. Porcari   CETENA, Italy
Ir. H. Scholte   Delft University of Technology,
                        The Netherlands

The Committee approved the course content and helped select the course lecturers whose notes form the material of these two books. The editors are grateful for their advice and assistance. The School was supported in part by COMETT funds which were obtained from the EC by the University Enterprise Training Partnership (Marine Science & Technology) which is administered through the Marine Technology Directorate Ltd., London. We are indebted to Mr. J. Grant, Secretary General of WEGEMT for his help in obtaining COMETT funding and in publicising the School.

The encouragement, support and assistance given to this venture by Professor W.G. Price and our other colleagues in the Department of Ship Science has been incalculable and valuable. We are grateful to them. Finally, we extend our thanks to Mrs. L. Cutler for her expertise and professionalism in undertaking the word processing and for patiently coping with the numerous edits, changes and amendments involved in preparing the camera-ready copy for the two books.

R.A. Shenoi, J.F. Wellicome
Southampton

LIST OF AUTHORS

- Professor H.G. Allen, University of Southampton, U.K. (Vol. 1)
- Dr. J. Benoit, Bureau Veritas, Paris, France. (Vol. 2)
- Mr. A. Bunney, Vosper Thornycroft, Southampton, U.K. (Vol. 2)
- Dr. D.W. Chalmers, Defence Research Agency, Haslar, U.K. (Vol. 1)
- Dr. P.T. Curtis, Defence Research Agency, Farnborough, U.K. (Vol. 1)
- Dr. R. Damonte, CETENA, Genova, Italy. (Vol. 1)
- Mr. A.R. Dodkins, Vosper Thornycroft, Southampton, U.K. (Vol. 2)
- Professor A.G. Gibson, University of Newcastle-upon-Tyne, U.K. (Vol. 2)
- Professor K. van Harten, Delft University of Technology, The Netherlands. (Vol. 1)
- Mr. G.L. Hawkins, University of Southampton, U.K. (Vol. 2)
- Dr. B. Hayman, Det Norske Veritas, Hovik, Norway. (Vol. 2)
- Mr. S-E. Hellbratt, Kockums AB, Karlskrona, Sweden. (Vol. 2)
- Mr. P. Krass, Schutz Werke, Selters, Germany. (Vol. 2)
- Mr. A. Marchant, AMTEC, Romsey, U.K. (Vol. 2)
- Professor G. Niederstadt, DLR, Braunschweig, Germany. (Vol. 1)
- Ir. A.H.J. Nijhof, Delft University of Technology, The Netherlands. (Vol. 1)
- Dr. G. Puccini, CETENA, Genova, Italy. (Vol. 1)
- Mr. R.J. Rymill, Lloyd's Register of Shipping, London, U.K. (Vol. 2)
- Ir. H.G. Scholte, Delft University of Technology, The Netherlands. (Vol. 2)
- Dr. R.A. Shenoi, University of Southampton, U.K. (Vols. 1 & 2)
- Dr. G.D. Sims, National Physical Laboratory, Teddington, U.K. (Vol. 1)
- Mr. P.J. Usher, Vosper Thornycroft, Southampton, U.K. (Vol. 1)
- Mr. I.E. Winkle, University of Glasgow, U.K. (Vol. 2)

1 A STRATEGIC OVERVIEW

1.1 INTRODUCTION TO PLASTICS

The earliest known plastic-moulding techniques were practised by Malayan natives in the early 1800s. They made utensils and artifacts from gum elastic (a vegetable material named gutta percha), softened in hot water and fashioned by hand.

The first synthetic plastic material produced was celluloid. Cellulose nitrate was first produced in 1835 by dissolving cellulosic products in nitric acid. Ten years later, a Swiss chemist by the name of Schönbein nitrated cellulose with sulphuric acid as a catalyst, and in 1854 an American scientist, J. Cutting, obtained patents for his process of gum camphor in collodion for photographic solutions. This marked the first use of camphor with cellulose nitrate. Evaporation of the photographic collodion by Alexander Parkes, an English scientist, produced a hard, elastic waterproof material that could be fashioned into useful articles.

Celluloid, although the first synthetic material, did not advance the moulded products industry as it could not be moulded. Shellacs and bituminous cold-mould plastics continued to dominate the industry until 1909, which marked the introduction of phenol-formaldehyde. Dr Bakeland is accredited with invention of phenol-formaldehyde resin, which was the first synthetic, mouldable plastic material. Since its debut, more than 50 distinct families of plastics have been invented with literally hundreds of variations of these materials. Table 1.1 summarises the major developments in the plastics field up to 1909.

Today, when reference is made to engineering plastics, those most frequently thought of are the acetals, nylons, phenolics, polycarbonates and fluorocarbons. The growth rate of plastics in industrial and commercial usage surpasses most conventional materials such as metals, rubber and ceramics.

High performance composites are formed by combining two or more homogeneous materials in order to achieve a balance of material properties that is superior to the properties of a single material. Increased strength, stiffness, fatigue life, fracture toughness,

1

Table 1.1. Chronological development of plastics.

1820	First rubber processing plant was built
1834	Liebig isolated melamine
1835	Pelouze nitrated cellulose
	Regnault developed vinyl chloride
1839	Goodyear introduced the vulcanization of rubber
1845	Bewley designed extruder for gutta percha
	Schonbein nitrated cellulose in sulphuric acid
1847	Berzelius developed the first polyester
1859	Butlerove made formaldehyde polymers
1865	Schuzenberger prepared acetylated cellulose
	Parkes's patented Parkesine process
1866	Berthelot synthesized styrene
1870	Hyatt patented basic celluloid
1872	Hyatt patented first plastics injection moulding machine
	Bayer observed reactions between phenols and aldehydes
	Baumann polymerized vinyl chloride
1873	Caspery and Tollens prepared various acrylate esters
1878	Hyatt developed first multicavity injection mould
1879	Cray patented first screw extruder
1880	Kahlbaum polymerized methylacrylate
1884	Holzen isolated urea-formaldehyde condensation products
	Chardonnet developed the first synthetic silk
1894	Cross and Bevan patented the first industrial process for manufacturing cellulose acetate
1899	Continuous cellulose nitrate film made by casting on a polished drum
	Spitteler and Kritsche patented casein plastic
1901	Smith studied alkyd resins in reaction of glycerol and phthalic anhydride
1905	Miles prepared secondary cellulose acetate
1909	Bakeland patented phenolic resins

environmental resistance, and reduced weight and manufacturing cost are some of the common reasons for developing high performance composites. The most common form of high performance composite material is the fibre reinforced plastic. High strength, high stiffness, low density fibres are embedded in a plastic matrix to form the fibre reinforced composite. Most of the strength and stiffness are provided by the reinforcing fibres; the matrix maintains fibre alignment and transfers load around broken fibres.

The development of fibre reinforced plastic composites during the last 20 years has been an explosion of technology. This was precipitated by the development of high strength glass and high modulus boron fibres in 1960 and the strong desire of the aerospace industry to improve the performance and reduce the weight of aircraft and space vehicles. In 1964 carbon fibres became available in research quantities and ultimately became the most widely used reinforcement in aerospace structural applications. In 1971 aramid (Kevlar) fibres became available commercially and are now being used extensively in automotive tyres, numerous aerospace structures and in some marine applications. Specific strength (strength-to-density ratio) and specific stiffness (stiffness-to-density ratio) for reinforcing fibres have continually been increased over the past 20 years. Specific strength and specific stiffness values for glass, Kevlar, boron and carbon fibres are compared with metals in figure 1.1 and are as high as 14 times the

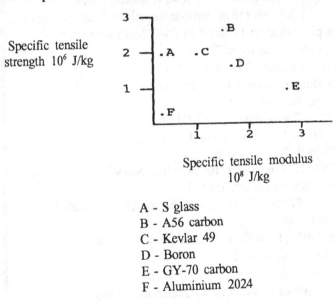

Specific tensile strength 10^6 J/kg

Specific tensile modulus 10^8 J/kg

A - S glass
B - A56 carbon
C - Kevlar 49
D - Boron
E - GY-70 carbon
F - Aluminium 2024

Figure 1.1. Comparisons of specific strength and stiffness.

specific strength and 10 times the specific stiffness of aluminium.

Table 1.2 shows that the first GRP boat hulls (polyester, single skin, and chopped strand mat) were produced in America at the end of the Second World War and were the forerunners of the new generation of GRP ships.

Table 1.2. Chronological development of first of class GRP ships.

1944 USN experience in GRP begins
1948 Minesweeping Launches built in GRP in US
1951 MOD(N) (UK) first involved 3 m boats
1953 First GRP boats at London Boat Show. GRP bridge fin in
      USS Halfbeak
1956 US 18 m minesweeper built (MSB23) Honeycomb Core
      Sandwich. 65 ft crew boat (Texas)
1959 Bow section fitted to US submarine
1960 Assault ship's pipes tested (UK)
1963 16 m high speed pick-up boat, single skin (USN)
      MOD(N) development of MCMV started (UK)
1964 Largest UK craft (16 m harbour launches) built
      22 m pilot boat in Holland in PVC foam sandwich
      20 m trawlers in Cape Town
      Halmatic single skin yachts (UK)
      25 m Russian river passenger boat
      100 ton Russian river tanker
1968 MCMV Test section tested (UK)
1970 47 m HMS WILTON started
1973 HMS WILTON completed
1974 VIKSTEN minehunter for Swedish Navy
1975 60 m HMS BRECON started
1977 47 m ERIDAN - French tripartite minehunter laid down
      Holland and Belgium follow
1978 50 m LERICI - Italian minehunter laid down
1979 BRECON completed
c.1980 Japanese fishing vessels up to 45 m
1982 31 m Rushcutter ordered for Australian Navy
      ERIDAN completed for French Navy
      LERICI completed for Italian Navy
1984 47 m LANDSORT class for Sweden in service
1986 13 m KISKI class minehunters for Finish Navy
1987 HMS SANDOWN single role minehunter started
1989 HMS SANDOWN completed
1991 BAMO class minehunter for French Navy in service

For about 20 years a variety of vessels of up to about 20 m were built in a number of different countries, but when in 1968 a section of a modern minehunter 2/3 full size was tested, progress towards a ship of 60 m was accelerated. HMS BRECON a mine countermeasures vessel for the Royal Navy entered service in 1979, and with the other twelve HUNT class, are to date the largest GRP hulled ships.

In order to gain service experience at sea with precisely the same material and type of construction as the projected HUNT class, HMS WILTON was converted from a wooden hulled minehunter to one of GRP, see figure 1.2. She has now (1993) had 20 years of normal usage without fault. Thus the pedigree of single skin, polyester resin woven roving E-glass hulls up to 60 m in length is firmly established.

Figure 1.2. HMS Wilton.

Double skinned hulls have had a more chequered history. Early examples became water logged, and in UK development trials it proved difficult to bond the shear carrying core material to the skins sufficiently well to withstand explosive shock loadings.

There are examples where cellular foam core material has reacted with bonding resin to produce a gas which has caused bubbles in the skins. However, there are now many examples of advanced composites in double skin configurations in high speed boats and in non-marine applications.

The early fears that fire would destroy a GRP structure have proved unfounded. Except in the case of thin skin, a GRP panel offers a remarkable fire barrier, and its low thermal conductivity prevents the spread of fire to adjacent areas. Similarly the problems with osmosis which occurred in early GRP small craft have not arisen in larger GRP ships.

As regards health hazards, over the forty or so years during which resins and fibres have been available on a commercial scale it is inconceivable now that asbestos was one of the first fibres to be used; fortunately not for long. Styrene is toxic, and its permitted level of acceptability differs in different countries, but no evidence has so far come to light that there are carcinogenic effects at the UK maximum level of 100 ppm for an eight hour per day exposure. Also, with correct protection, the incidence of dermatitis, which was a problem in early days, is now very rare.

1.2 MARINE APPLICATIONS

Sailboards, dinghies, yachts, lifeboats, power boats, fishing boats, workboats, patrol boats and minehunters, such as HMS Sandown (see figure 1.3), are now almost universally made of GRP as are sonar

Figure 1.3. HMS Sandown.

domes and the structure of some submarines outside the pressure hull. Surface effect ships or hovercraft and small catamaran ferries commonly use GRP, and the pressure hulls of a number of deep submersibles have also been constructed in the material. Some high performance power boats and some Norwegian multi-purpose survey vessels are using advanced composites. Filament wound pipes are used for ballast systems in some ships. A few offshore oil platforms have GRP fire-resistant helicopter decks. Kevlar has been used as armour protection against small arms fire in a variety of ships.

These marine applications, in addition to the widespread use of composites in aircraft (the European Fighter Aircraft, EFA, is designed to have 70% of its airframe in carbon fibre reinforced plastic) gain the advantage of weight saving over traditional materials and thereby operate with improved performance.

The absence of corrosion, rot and marine borers reduces maintenance costs, and in most of the current ship applications more than offsets the greater first cost of GRP materials.

Applications which are being actively considered include:

1. The superstructures of frigates
2. Protective shells for vital components in offshore rigs (post Piper Alpha)
3. Accommodation modules in offshore rigs
4. Sheaths for the legs of rigs to prevent fire damage from a seaborne fire
5. Pipes and tanks on offshore rigs
6. SWATH, submarine and submersible hulls

Currently it is believed that no serious work is being done on the design of FRP hulls in excess of 60 m, but undoubtedly there will be some.

A further area of possible application is in the civil engineering side of ports and docks where originally granite and then cast iron, but now mainly reinforced concrete is used. It would probably be economic to make dock gates and caissons from FRP.

1.3 MATERIALS TECHNOLOGY

E-glass and polyester resin are the components in the vast majority of today's marine composites. Molten glass extruded into filaments of about 0.01 mm develops its full strength by virtue of maximising the molecular bonds in one direction. When set to shape in a matrix of well adhering resin the resulting composite will bear significant loads in the direction(s) of the reinforcement fibres but with only modest strength orthogonally.

Other reinforcements and resins are available which offer higher specific strengths and stiffnesses at increased cost. The resins, which have limited shelf lives, are mixed immediately before use with between 1% and 6% by volume of catalyst/accelerators.

The catalyst/accelerators cure the resin by reacting with styrene (a monomer in polyester resins) to give a "gel time" of some hours during which the composite can be moulded to shape. During this time styrene vapour is emitted, but if caused to evaporate too rapidly, for example by blowing high velocity air across the laminate, the cure of the resin will be impaired. Heat is generated by exotherm during the chemical reaction, limiting the amount of composite (number of layers of laminate) that can be superimposed at one time.

As the materials technology is governed by chemical reaction it is essential that only the specified constituents be incorporated. Cleanliness, including dust extraction, is therefore necessary, but not dehumidification. A minimum temperature of 15°C whilst laminating and curing is normal.

Fibres of glass (E, R or S type), or Aramid, Boron or Carbon would be unmanageable unless worked into a fabric. From a range embracing the simplest random chopped strand mat to special weaves giving improved through the thickness strength, or better drape, woven and knitted fabrics are increasingly being used to develop greater strength and greater productivity. Strength is increased by reducing the crimp of the fibres. Thermoset resins may be polyester, epoxy, phenolic or vinyl ester and the main issues cost, adhesion, shock loading particularly where the resin is carrying shear forces, degradation in water and smoke emission. Weight, strength and stiffness can also differ by about 20% between these matrices.

A resin to fibre ratio by weight of 50:50 has been normal with polyester/glass composites to provide reliable adhesion.

This can be reduced to 30:70 (resin to fibre) with the associated weight-saving and thinner laminate, if a modern process, such as SCRIMP (see section 1.5 below), of vacuum bag and resin distribution is employed.

The structural efficiency of many applications is best achieved with a sandwich configuration. A core material bonded between two layers of FRP enables flexural loads to be carried provided the core, which is in shear, is bonded adequately to the inaccessible inside surfaces. End grain balsa wood has been used widely as a core material, as have aluminium honeycomb, and polystyrene, polyvinyl chloride (PVC), polyurethane and acrylic foams.

To enhance the already good resistance to fire exhibited by GRP it has been found that a double skinned panel with a core of dense insulating foam is capable of withstanding a hydrocarbon fire for two hours. Temperatures of 1000°C have not burned through these panels which have also been designed to withstand the blast pressures associated with a major fire on an oil rig.

On composites in general it is possible to pigment the resins to improve appearance without the need for painting, but this is not widely practised on important structural work because it prevents voids, blemishes and other faults in production being readily seen.

1.4 DESIGN TECHNOLOGY

The use of advanced composites can lead to weight-savings compared with aluminium though at greater cost whilst the more common polyester/glass mixtures are about the same weight as aluminium, which is itself about half the weight of steel for most ship structures.

Besides the improved performance or greater payloads to be achieved by lighter structures, the life-cycle cost reduction brought about by the reduced maintenance of GRP now appears to be significant.

Not all owners - whether navies or commercial operators, are willing to base decisions on through life costs, largely because of their unreliability in the past. Efforts should be made to so collate running costs as to identify the net savings to be made by higher specifications at the time of purchase.

As with each new step in engineering it is highly desirable that a structure to be made of FRP should be designed to best utilise the characteristics of the material, rather than merely substituting one material for another in an existing design. It is equally important that the designer fully understands the limitations and scope of the production processes.

Some of the pitfalls encountered by designers more accustomed to metallic materials are:

- **Peeling**: Joints commonly fail as a result of laminations running from one plane to another, e.g. a shell to frame connection, pulling away from the member to which it is attached by resin bonding in tension - as with peeling an orange.
- **Shockloading**: Both underwater explosive loading and ship slamming can impose extremely high pressures over small areas of the outer bottom. Different resins have different strengths under these conditions.
- **Tank boundaries**: The low modulus of most composites

compared with steel causes rotation at the boundaries of tanks under pressure, which can lead to microscopic failures and leakage.

- **Through-the-thickness-strength**: Though most calculations are done as for an isotropic material it must not be forgotten that FRP is anisotropic. Tension perpendicular to a laminate can cause delamination before failure at the bonded interface, depending on the areas involved.
- **Impact loads**: The light weight of FRP contributes to higher velocities when impacts are delivered, whilst the lower modulus allows greater deflections to absorb the energy.
- **Creep**: Long term static loads are generally accounted for by increasing the margin of safety, though not enough literature is available on the subject.
- **Design criteria**: Progress with the adoption of GRP as a structural material has been, and still is impeded by the absence of reliable design criteria. The Classification Societies have issued some rules for the most common composites but there are cases where these are conservative, and others e.g. core materials, where more work needs to be done for designers to feel comfortable.
- **Slinging**: Movement of large FRP structures by crane can cause damage. Special lifting frames may be necessary.

There are many other design considerations beyond structural strength. Electro-magnetic radiation from, for example, concentrated radar transmitters penetrates FRP structures. Personnel and sensitive electronic equipment inside such structures may need screening protection. The electrical insulating properties of FRP might allow the build up of an electrostatic charge on metallic pipe fittings through which a liquid way is pumped at high velocity. Suitable earthing strips may need to be fitted.

Electrolytic corrosion of metal hull fittings widely spaced in the electro chemical series can be more serious in an FRP vessel. Potable water tanks do not suffer from styrene contamination and need not be painted. Chafing of FRP must be avoided because it can lead to the exposure of large areas of unprotected fibres; osmosis can be started at such positions.

1.5 PRODUCTION TECHNOLOGY
Manufacturing a material and a product in one process in unusual, and

it has probably damaged the reputation of the composites industry when one or the other was less good than it should have been. The fact too that no expensive facilities are necessary to produce a GRP dinghy by spraying chopped strand mat, has meant that a large proportion of the population believes that all composite structures are crudely made "under the arches". Chopped strand mat dinghies are known to have disintegrated when burned; ergo, all GRP burns away to a cinder.

Using existing processes, the production facilities necessary to build high quality thermoset composites for marine use have cost rather more than mechanical engineering workshops of the same size. Buxton certified electrics should be installed and hand tools should be air operated. The fork lift trucks used must meet a stringent spark free specification; a sniffer monitoring system which continuously records styrene levels is a highly desirable safety precaution. Special ventilation is required because of the minimum temperature needed and, coupled with the air changes to keep styrene levels satisfactory, a heat exchanger in the ventilation system becomes economic.

It may prove possible to relax some of these requirements with the advent of new processes which contain the emission of styrene.

Continuous control of the quality of the laminate is paramount. This responsibility is best invested in the laminators themselves, bolstered by full time inspectors who are relatively few in number but with power to insist on remedial measures when necessary. The most common defects are the inclusion of air bubbles between laminates and the failure to 'wet-out' the fabric sufficiently to ensure complete adhesion of the fibres and resin. Care must be taken to keep a dust-free atmosphere.

About three quarters of the marine market for composites goes into the production of leisure craft which in the main are formed by sprayed E-glass chopped strand mat and unsaturated polyester resin. Woven rovings where used, are incorporated by hand lay-up.

The remaining 25%, i.e. the commercial market, is made up of about 10% naval, 10% safety vessels and the balancing 5% is workboats, where the bulk of the laminate is made with woven rovings. In this area there have been attempts to improve productivity and to get more consistent quality from using resin and fabric dispensers, and more recently prepregs. Success with dispensers has been elusive to date mainly because the workloads have not been sufficiently high or consistent to maintain continuous production. Dispensers require meticulous cleaning if worked intermittently.

Developments with vacuum bags, whereby atmospheric pressure is used to consolidate laminates, have been ongoing, but a breakthrough was made in America three years ago by the Seemann Corporation whereby dry fabric covered with a membrane which incorporates a distribution medium is impregnated in situ under atmospheric pressure.

The SCRIMP process as it known (Seemann Composites Resin Injection Process), produces high quality void-free laminate, but with the important added advantage that much lower resin to reinforcement ratios are possible than hitherto. Strength properties of E-glass/polyester laminates have been doubled compared with 50/50 glass/resin hand lay-up, by achieving 70% glass with the SCRIMP process. Reinforcements of 80% have been achieved with indirectional material. Because the process is enclosed, the emission of styrene is localised and reduced, as less resin is employed. It has also proved possible to build a thicker laminate, up to 50 mm, wet-on-wet with acceptable exotherm, because the proportion of resin-to-glass is low. In general the SCRIMP enables an equivalent strength to be achieved with thinner and therefore lighter structure. Impact resistance has not yet been determined rigorously. Vosper Thornycroft have acquired an exclusive (in the UK and EC) licence to manufacture military and paramilitary craft of more than 15 m but excluding hovercraft, using this patented process, and to license others.

The shipyard layout required for the building of GRP ships must incorporate weather proof buildings for the production of panels and minor mouldings, and the complete hull, and dry storage for fibre. A mixing room with safety precautions against fire is required for ready use resin and catalyst/accelerators. The temperature in all these spaces should not fall below 15°C. They should therefore be generously insulated.

Ideally another large building is required in which three dimensional units of interior or exterior structure can be assembled and fitted out before being transferred to the single hull moulding. When bulkheads, flats and decks have been incorporated and underwater work completed, and as much upper deck work as possible finished off, the ship can take to the water uncovered.

The hull mould - necessarily a female mould for large ships which it is impracticable to turn over, may be made from wood, steel, aluminium or GRP and must be capable of being removed from the moulded hull without the hull being disturbed. This can be done by removing sections of the mould in sequence and transferring the hull weight to keel or other blocks on the building berth. It is possible if the berth is wide enough, to remove one whole side of the mould and then pull the ship across the berth transversely leaving the other half of the mould in place.

Cutting panels to shape is commonly done by power sawing which produces dust. For large scale production, abrasive grit can be used in computer controlled machines operating in a water bed, in much the same way as plasma arc metal cutting.

Thermoplastic pultrusions are gradually expanding their applications. These are produced by specialist suppliers and are being used amongst other things for gratings and working platforms in offshore rigs where previously metal decking had been used.

Lifting frames to enable panels to be supported at appropriate spacings to prevent excessive bending are necessary for handling large panels of GRP.

The Control of Substances Hazardous to Health (COSHH) regulations apply to the storage and handling of resins, solvents and accelerators, and the UK Health and Safety Executives are responsible for checking safety matters in general, but particularly in the case of composites, the exposure of employees to styrene vapour.

1.6 THE FUTURE

Twenty years of service experience at sea with HMS WILTON and much longer with some other applications, backed up by thousands of boat hulls and about one hundred ships hulls in excess of 45 m, is proof that the thermoset E-glass/polyester composite is a very satisfactory material for the marine environment, and for small ships and boats far superior to metal or wood. Those hulls which have been manufactured to a recognised standard of quality are showing little or no signs of degradation well beyond the end of what would have been regarded as a normal life span for wooden vessels. Life cycle costs for craft made of this material are demonstrably low. Fatigue has not been a problem.

With accepted safety measures in place, no adverse medical reports are known to have been made on production personnel, beyond a small proportion of people who has suffered with dermatitis from handling glass fibres.

Most resins and certainly the polyesters burn readily and reference has been made to the need for stringent precaution in production facilities. Once the resin has set, it is more difficult to ignite but its surface will inevitably burn if exposed to high temperatures as in a ship fire. Tests and full scale experience have, however, demonstrated that the glass, particularly if in the form of a woven roving, provides high levels of insulation and that not only does the fire not penetrate the structure very far, particularly if the structure is fairly thick, but the temperature on the remote side is so low that the fire does not spread to adjacent compartments.

Thus a GRP ship, though made of flammable material, will often be safer against fire hazard than a steel ship because of the enormous difference in the insulation value of the material.

Composites offer the prospect of weight saving, fire resistance and low maintenance as well as being non magnetic. Weight saving in the structure of a slow ship is not of great importance and many operators of commercial vessels have learned to live with corrosion, so it is unlikely that small or large merchantmen will be built of composites. High speed, high performance vessels such as the craft shown in figure 1.4, however, may be. Flexural stiffness and cost prevents long vessels from being formed of composites today but advanced hybrids may offer a solution, though an FRP QE2 is most unlikely.

Figure 1.4. The 'Viper'.

Ferries could certainly benefit. Naval corvettes of up to 100 m and say 1500 tonnes will probably become more numerous in the years ahead and they will benefit from lighter hulls which do not corrode. Patrol vessels, workboats, and large yachts still being built in aluminium (or steel) will gradually change over to composites to reduce through life costs and gain additional safety from fire. The non-magnetic steel minehunter's days have reached a plateau, while composite materials technology is advancing rapidly.

There is an enormous market to be won in the offshore oil industry where fire and blast protection of emergency shut down valves, personnel shelters, accommodation modules and piping and decking can derive benefits from the fire barrier that FRP provides and the weight-saving over steel that it brings.

There is clearly an ongoing market in leisure, and new applications to be made in the commercial and naval fields. The aerospace industry will continue to develop its own applications and also to conduct R&D into thermoplastics and the processes necessary for their use. R&D for marine applications on a large scale where confidence is lacking would probably best be concentrated on thermosets which have established a pedigree of reliability. There is still much to do to enable designers to choose between polyester and epoxy, phenolic or vinyl ester resins, with a range of reinforcements, to provide strength and stiffness cost effectively.

The racing boat fraternity are a valuable source of R&D, if sometimes ad hoc, and will probably make progress with carbon and boron fibres and possibly with thermoplastics.

There are many research projects in hand today. They are funded by the international oil companies, material suppliers, major contractors, and for those in the UK, the Department of Energy, the Ministry of Defense, the Science and Engineering Research Council, the EC and doubtless many others. The principal target of such studies ought to focus on regularising design and manufacturing standards, thereby improving maritime safety and cost-effectiveness of boats, ships and rigs.

2 BACKGROUND TO MATERIALS SCIENCE

2.1 CLARIFICATION OF WHAT ARE COMPOSITES

Fibre reinforced polymers are highly qualified materials, suitable especially for innovative developments in all technical domains because of their potential for weight-savings and reduction of manufacturing costs.

The wide variety of composites and their variation in mechanical properties requires a good knowledge of the constituents like fibres and polymers together with their influence on the behaviour of the composite material. Before going into details about fibres and resins (or polymers) some cases will be discussed which demonstrate where composite material can be used with benefit.

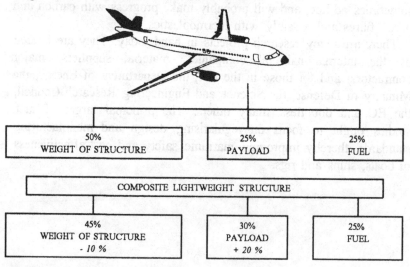

Figure 2.1. Economical gain by using lightweight materials in aircraft.

Aircraft engineers are always interested in saving weight. This becomes obvious on looking closely to the operation costs of a commercial plane [1,2].

According to figure 2.1 the total weight of an aircraft can be divided into the weight of the frame, the payload and the weight of fuel. Most weight is concentrated in the airframe, and if weight-savings are possible the percentage increase in payload would be almost twice the percentage saving in airframe weight, therefore improving the economy of the aircraft significantly.

One of the most impressive examples for well designed lightweight parts in aircraft is the horizontal stabiliser of the airbus, seen in figure 2.2. The weight-saving was about 20%.

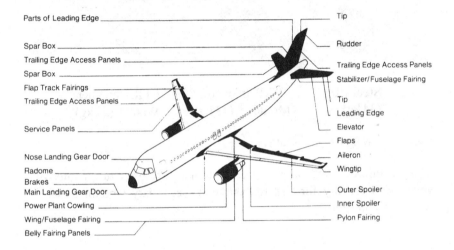

Figure 2.2. Carbon and aramid FRP components (in black).

Another very obvious reason for weight-savings are given for all parts which are sometimes highly accelerated or retarded. This is based on Newton's law. Sir Isaac Newton (1642-1727) formulated the law of the inertness of mass:

Force = Mass x Acceleration

In a technical sense, it means that reduction of mass will directly reduce forces or energy consumption as the mass is accelerated or retarded.

Good examples are racing cars, rockets, pistons and connecting rods. For these applications composite materials are of high technical benefit if they are able to reduce the weight of the construction. Figure 2.3 demonstrates the weight reduction of a connecting rod achieved by redesigning with composite materials.

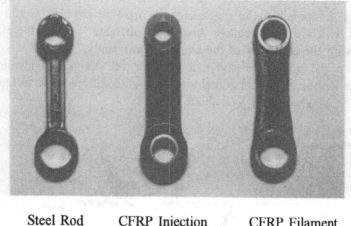

Steel Rod   CFRP Injection   CFRP Filament
(34 kg)   Moulded (13 kg)   Wound (27 kg)

Figure 2.3. Weight reduction of connecting rods.

Many reinforced and unreinforced polymers are used in automotive vehicles, especially for cost reduction in "secondary structures" as seen in figure 2.4.

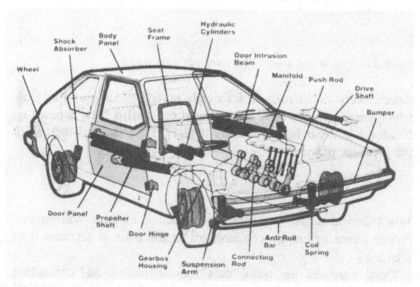

Figure 2.4. Secondary structures in an automobile.

2.2 FIBRE SELECTION

A wide variety of fibre types is available for reinforcing polymers. It is up to the engineer to select the most suitable type for his application using one or several optimising criteria. Most important for application of composite materials is the reduction of weight by increasing strength and stiffness. Therefore the design engineer should not only refer to strength and stiffness of a fibre type but also to density. Both strength or stiffness and density are included in the so-called specific strength or stiffness. Therefore the most suitable criteria for fibre selection are high values of:

$$\text{SPECIFIC STRENGTH} : \sigma / \rho \ g \ [\text{km}]$$
$$\text{SPECIFIC STIFFNESS} : E / \rho \ g \ [\text{km}]$$

Figure 2.5 shows a comparison of some materials, which are reinforced with different fibre reinforcements.

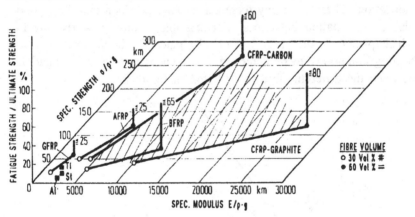

Figure 2.5. Specific strength and specific modulus of fibre reinforced materials.

The first fibres used, were glass fibres, which are relatively cheap and able to raise the specific strength remarkably compared to metallic materials like steel, aluminium and titanium. However glass fibres offer no benefit in specific stiffness, which is about the stiffness of aluminium only.

For achieving higher stiffness, boron fibres were developed. They are of similar density to glass fibres but of much higher stiffness, approximately double that of steel. These fibres are especially useful for composite structures with stability problems.

Unfortunately boron fibres are very expensive in terms of the costs of basic materials and the cost of the production process. They are also difficult to handle because of their small fibre diameter of about 100 microns. Therefore they were progressively replaced by carbon fibres. Carbon fibres offer the most impressive value in specific strength and specific stiffness. Unlike glass and boron fibres a wide variety of carbon fibres is available. They are crudely divided into fibres with high strength (HT) and fibres with high modulus (HM). This will be discussed later in more detail.

Other highly important fibres are the aromatic polyamids or "aramids". These are produced by pultrusion of highly crystalline polyamids and derive their good mechanical behaviour from the strong alignment of the crystals during the production process. Since the densities of aramids are very low (ρ = 1.45 g/cm3), these fibres are of good potential in terms of specific strength and stiffness.

The stress-strain diagram is quite informative. It provides data about elastic or inelastic behaviour as well as about fracture elongation. There are many technical problems which call for a good plastic deformation behaviour. For example, a multi-row riveting in which plastic deformation has to take place for homogeneous load bearing conditions to be achieved. Differences in fracture elongation between the different fibres are plotted in figure 2.6.

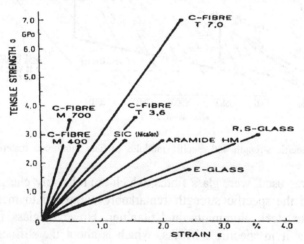

Figure 2.6. Stress-strain diagram of different fibres.

Carbon fibres are fully elastic and of high modulus and high strength, but do have low fracture elongation (except the carbon fibre T 7.0 which has recently been developed and is only available on laboratory scale). Aramids and glass fibres, especially high strength glass fibres, are of higher ductility and somewhat viscoelastic. So they are suitable for localised loading problems.

Synthetic fibres, like aramids, are more suitable if impact is an important design criteria. Significant differences in terms of impact behaviour are found between the commercially available fibres. Results are shown in figure 2.7, where the notch toughness (measured by the Charpy test) is compared between different carbon and aramid-laminates.

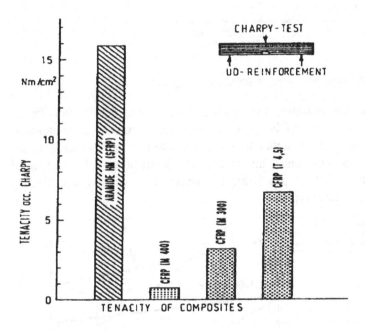

Figure 2.7. Comparison of energy consumption of different UD-FRP.

Some materials, especially if they are reinforced with HM-carbon fibres are extremely brittle and cannot be used with benefit where energy absorption is required. Synthetic fibres, like aramids, are vastly superior as energy absorbers. The application of aramids in safety vests used as ballistic armour for policemen or for safety belts in automobiles is well-known.

On the other hand aramids are organic fibres which are viscoelastic and therefore have the tendency to creep if they are continuously and highly loaded. This is shown in figure 2.8, where measurements are plotted from long time loading under laboratory conditions and compared with those from carbon fibre reinforcements.

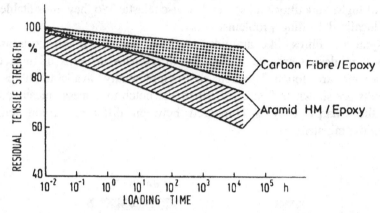

Figure 2.8. Long term strength (creep) of composites.

The plot demonstrates, that aramids creep more than carbon fibres. The degradation will be even worse if other outside environmental conditions, such as higher temperature, moisture or UV-radiation are present. In this case an additional degradation of mechanical properties will take place due to ageing. This will be discussed in more detail later on.

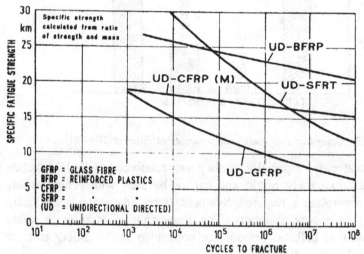

Figure 2.9. Specific fatigue strength of different composite materials.

In order to estimate the behaviour under dynamic loads, the S-N-diagram is quite useful. Figure 2.9 shows such a graph for boron fibres, aramid fibre, carbon fibre and glass fibre laminates, all with unidirectional fibre-orientation. As a design criterion, absolute fatigue strength is not as interesting as the loss of strength during the component's lifetime or the proposed number of loading cycles. In this sense there are significant differences between different advanced composites. Table 2.1, for example, gives typical values of strength ratios for various composites.

Aramids and glass fibres are most sensitive to fatigue. Carbon and boron fibres are much less influenced by fatigue loads as seen in the graph.

Table 2.1. Typical values of specific fatigue strength.

carbon fibre reinforced laminate (CFRP) $\sigma(_{10}{}^{8}) / \sigma(_{10}{}^{4}) = 0,83$

boron fibre reinforced laminate (BFRP) $\sigma(_{10}{}^{8}) / \sigma(_{10}{}^{4}) = 0,80$

Aramid fibre reinforced laminate (SFRP) $\sigma(_{10}{}^{8}) / \sigma(_{10}{}^{4}) = 0,40$

glass fibre reinforced laminate (GFRP) $\sigma(_{10}{}^{8}) / \sigma(_{10}{}^{4}) = 0,45$

When assessing the thermal behaviour of composites, the type of matrix is highly important. However there are big differences in the temperature resistance of the fibre reinforcements, measured in an inert atmosphere or in an oxygen environment. Figure 2.10 gives an

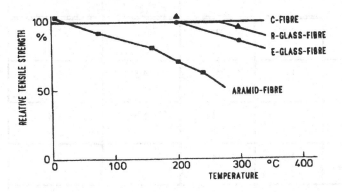

Figure 2.10. Influence of temperature on tensile strength.

overview of the onset of loss of strength by the different fibres with increasing temperature.

Aramid fibres will start creeping at a little above room temperature and therefore incur a heavy loss of strength due to increasing temperature. Degradation of E-glass fibre is found above 200°C followed by R- and S-glass fibre which retains strength up to 300°C. The carbon fibres are consistently strong up to 1000°C. Within carbon fibres, the HM-type fibre is the most thermally resistant one.

Not only the thermal resistance but also the mechanical characteristics are quite different for different carbon fibre types. A wide variety of carbon fibre types is produced by the industry, offering different stiffness, strength and breaking elongation characteristics. Table 2.2 lists the most common types without their

Table 2.2. Carbon fibres for advanced composite structures.				
FIBRE CODE	DENSITY g . cm^{-3}	YOUNG'S MODULUS GPa	STRENGTH GPa	FRACTURE ELONGATION %
M 300 (M 20)[‡]	1.8	300	4.0	1.3
M 400 (M 40)[‡]	1.8	400	2.7	0.6
M 500 (M 50)[‡]	1.9	500	2.4	0.5
M 700 (Pitch Fibre)	2.17	700	3.2	0.46
T 3.6 (T 300)[‡]	1.7	230	3.6	1.5
T 4.5 (T 400)[‡]	1.8	250	4.5	1.8
T 5.6 (T 800)[‡]	1.8	300	5.6	1.8
T 7.0 (T 1000)[‡]	1.8	300	7.0	2.3
M: MODULUS T: TENACITY	[‡] Trade name of Toray Industries, INC.			

trade names. The stiff ones have been marked with their moduli (M) and the strong and tough ones with T (for "Tenacity", or toughness, and strength). Pitch-fibres have not been listed. According to this scheme they are HM-fibres with extremely high modulus of elasticity.

Much more illustrative than table 2.2 is a graph showing the different fibres according to their tensile strength and modulus of elasticity. Figure 2.11 shows all products which are presently available on the market.

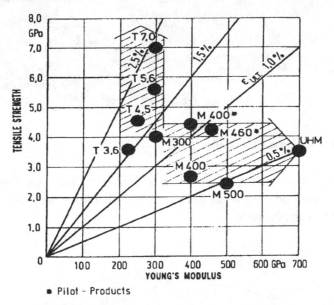

• Pilot – Products

Figure 2.11. Tendency of improving carbon fibres.

The trend in the development of fibres is towards fibres with higher stiffness and breaking elongation because these suit the requirements of the practical engineer, who prefers ductile materials capable of load transfer.

The question arises as to whether so many carbon fibres are desirable. This question has to be answered with "Yes" because the variety of carbon fibres enables the engineer to optimise materials for new applications. An example is the noseprobe on an airplane, see figure 2.12. This detector has a metallic sensor at the tip, which induces vibrations. By choosing a cabon fibre with high modulus of elasticity the specific resonance frequency can be moved to an above-critical level.

A further typical example of optimised fibre selection is a

Figure 2.12.   Noseprobe of an airplane [3].

fast-moving flywheel made of carbon and aramid fibres, see figure 2.13. Damping and elongation have been the criteria for the choice of fibres for this application.

Figure 2.13.   Fast moving flywheel.

For several applications mixtures of different fibres might be advantageous. However, the mixture of fibres does not always result in an improvement of mechanical characteristics. The main reason for this phenomenon is incompatibility of the breaking elongations.

An example of the stress-strain behaviour of a hybrid laminate is seen in figure 2.14. The modulus of the laminate can be calculated by using the rule-of-mixtures according to the volume of each fibre type. However, the strength is limited by the fibre with the smaller breaking elongation. These are mostly carbon fibres. Due to the partial carbon fibre breakages in this hybrid-composite, energy absorption results in a stress rearrangement to the glass fibres. This will increase the energy absorption of this hybrid-composite.

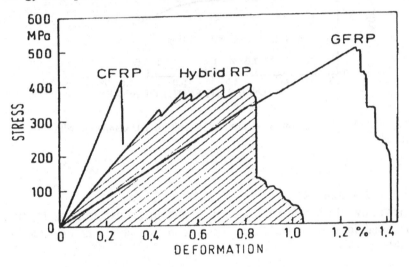

Figure 2.14. Comparison of deformation behaviour of carbon fibre, glass fibre and hybrid laminate.

Composites with newly developed fibres are often offered as especially efficient products. But the benefit is sometimes very limited. In fact high strength fibres such as T 800 or T 1000 have a higher tensile strength because they are manufactured as very thin elementary filaments (diameter of 5 μm). They achieve higher tensile strength, but buckle very early under compressive loads so that bending and buckling strength is made worse.

When plotting the strength of unidirectional composites with the fibres T 300, T 400 and T 800 over the real fibre strength, bending and compressive strengths show a remarkable degradation with increasing fibre tensile strength as seen in figure 2.15. The result demonstrates, that structures with critical compressive or bending strength cannot be improved by the use of high strength carbon fibres. This is normally the case for engineering structures.

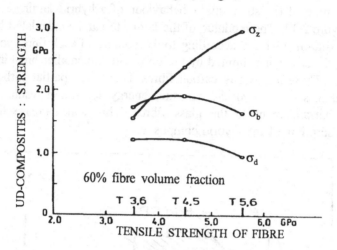

Figure 2.15. Comparison of mechanical benefit of UD-composite with different fibres.

2.3 MATRICES

Generally, there is a variety of polymers to choose from as matrix material for composites. Some typical common polymers are listed in table 2.3.

Table 2.3. Usual polymers for matrix materials.
1. Unsaturated Polyester
2. Vinylester Resin
3. Epoxy Resin
4. Phenolic Resin
5. Polyimides
6. Thermoplastic Resin: PA, PC, PEEK, LCP
7. Elastomeric Resin: Polyurethane

For normal industrial applications unsaturated polyester resin is used; for applications strongly exposed to water vapour vinyl ester-resins are chosen. In airplane construction epoxy-resins are favoured because of their better dynamic characteristics. In the case of similar high requirement for inflammability, phenolic-resins. are preferred. They are becoming more important as their curing process is by now almost the same as for the polymerised and polyadducted polymers.

If high temperature resistance is required (over 200°C) epoxy resins are no longer useful. Resins with higher thermal stability but with similar curing technology are polyamides. These are much more expensive, so that they are nowadays only used for space applications. Thermoplastic matrices are being utilised more often because the manufacturing of proper semifinished material has been extensively improved. The most widely used polymeric matrices - such as polyethylene, polypropylene and polyamide and other qualified polymers have to be mentioned. The main benefit of the thermoplastics is the higher damage tolerance, which is obvious from the stress-strain diagram. Attention should be given to the high plastic deformation capability which can be increased by additional moisture pick up. This can be seen in figure 2.16, where a stress-strain curve of PEEK is plotted.

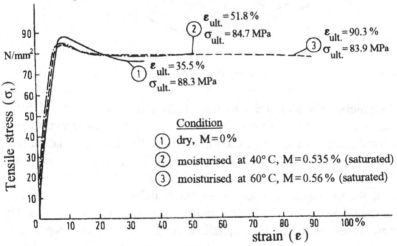

Figure 2.16. Elongation of PEEK as function of moisture pick-up.

2.3.1 Viscoelasticity

It is important for the engineer to know that - unlike metals - all polymeric matrices have a viscoelastic deformation behaviour. Due to this time-dependent stress and strain the diagram has to have a third dimension, which is the time, illustrated in figure 2.17. Normally the three-dimensional diagram is projected onto the stress-strain plane, which produces an isochronic stress-strain diagram. The projection onto the time-strain plane gives the creeping diagram. The time-dependent deformation is called viscoelastic deformation. It is rheologically described by two deformation-models namely the Hooke spring and the Newton dashpot.

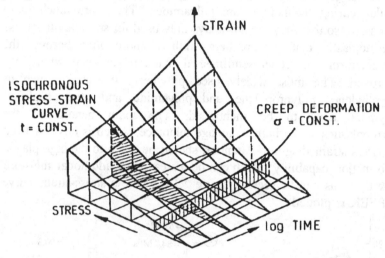

Figure 2.17. 3-D Isochronous stress-strain diagram.

Polymeric matrices are - in their behaviour - a mixture of both basic models, which means that as well as an elastic reaction the time-influence, or the creep, according to the Newton deformation has to be taken into account. In this respect the molecular structure is of importance. Generally, linear and branched chain molecules - as they are in thermoplastic matrices - are more influenced by the loading time than the cross linked thermosetting matrices. However, within the thermoplastic matrices the crystalline areas are less time-dependent than the amorphous ones, so that the crystallisation rate is highly significant to the degree of viscoelasticity.

As is often explained, it is not enough to describe a polymeric material by a "snapshot" of the stress-strain curve. Stress-strain curves have to be developed for all relevant loading rates. An example of a polyester-laminate with glass fibre mats is given in figure 2.18.

Although the inorganic fibre reinforcement, such as for example glass or carbon fibres, are fully elastic, reinforced laminates may also

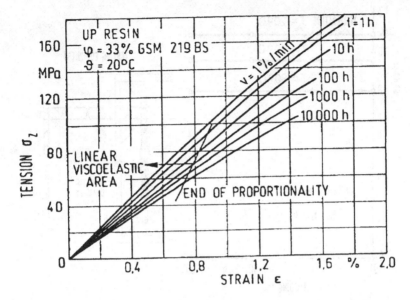

Figure 2.18. Stress-strain curves for GFRP with chopped strand mat reinforcement.

have a viscoelastic behaviour. This is due to the various micro-mechanisms working within the composite. In chopped strand mat laminates the propagation of load (loadpath) is always via the polymeric areas, so that its viscoelastic characteristic is reflected in the composite material properties.

This is also relevant for fabric reinforcements, where the weave generates stress in the resin. Absolute elastic characteristics can only be obtained using unidirectional straight fibre reinforcements, such as for example the CFRP-prepreg. This is shown in figure 2.19. Creep tests at 20°C, 40°C and 80°C, see figure 2.19, have not shown any viscoelastic deformations in the UD-prepreg-laminates in the fibre direction. Even at higher temperatures up to 180°C it has not been possible to find any creep effects. These "fibre-controlled" characteristics are not found with woven fabric reinforcements.
With increasing temperature the length of the carbon fibre reinforced composites will decrease, because of the negative coefficient of thermal expansion of the fibres.

The viscoelasticity of some composite materials is obviously very important in the case of impact loadings, see figure 2.20. A high loading rate seems to generate high stiffness values and also an increase in breaking elongation. The polymeric matrix solidifies like

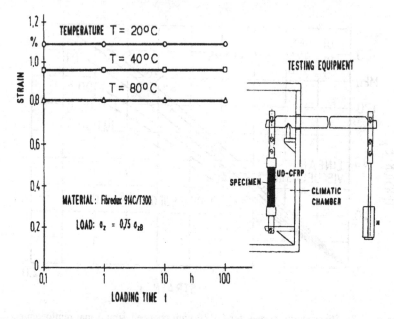

Figure 2.19. Creep behaviour of a UD-CFRP-laminate upto 100 hours.

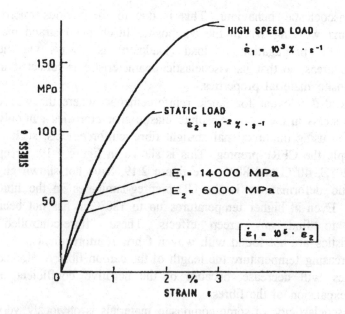

Figure 2.20. Influence of loading speed on stress-strain behaviour of GFRP.

in a shock second.

2.3.2 Damping

Materials with small viscoelasticity, also have little structural
damping. Damping may be an important design criterion for cases of
dynamic structural loading. It is therefore necessary to know how
prepreg-laminates generate viscoelastic behaviour and therefore
damping. Damping measurements on dynamically loaded composites
with directions of the fibre reinforcements at +/- 45° (for a high
matrix-load) allow a good comparison between polymeric matrices,
see figure 2.21.

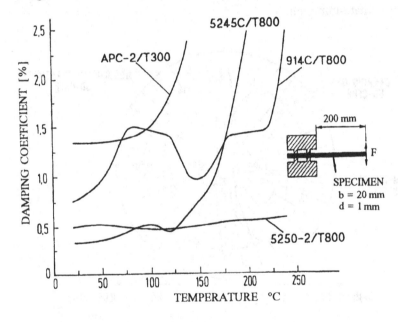

Figure 2.21. Temperature dependent damping behaviour of different kinds of CFRP.

Although the fibre reinforcements were of the same type and
quantity (up to APC-2) there are significant differences in the curves
with regard to the damping behaviour attributable to the differing
matrix types. Different polymeric matrices - for example thermoplastic
and thermosetting matrices - and also different cross-link-rates can be
recognised by the various maxima of damping with increasing
temperature. Strongly oscillating curves lead to the conclusion that
those are not pure resins, but mixed ones, with mixed-polymeric
matrices. Remarkably higher damping has been measured on fibre
reinforced thermoplastic matrices. This result shows the influence of

the matrix materials on damping and viscoelasticity and also that damping measurements as temperature-functions might be useful for quality tests.

By varying the angle of the different fibre-orientations in respect to the load axis, polar damping diagrams are obtained, see figure 2.22. Such a diagram shows the strong influence of the fibre-orientation and the important fact that the damping is very small when the modulus of elasticity has a maximum. This occurs when the load axis is aligned with the fibre-orientation. Maximum damping has been measured at an angle of 30° to the fibre-orientation. The angle of below 90° to the fibre-orientation might be called an intermediate-minimum.

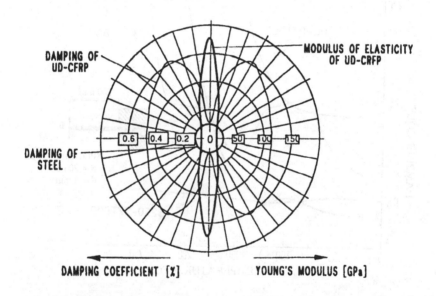

Figure 2.22. Polar diagram for stiffness and damping of UD-CFRP [4].

Also important for the engineer is the question of the damping characteristic of multi-layer-composites. First tests on a bidirectional-reinforced laminate are illustrated in figure 2.23. The dominating influence of the fibre reinforcement is clear to see. The damping maxima can be found again at 30° of the fibre-angle to the main-fibre-layer.

In the fibre direction there is the same damping behaviour as in steel. It is therefore understandable that "fibre-controlled" constructions show bad damping under dynamical load.

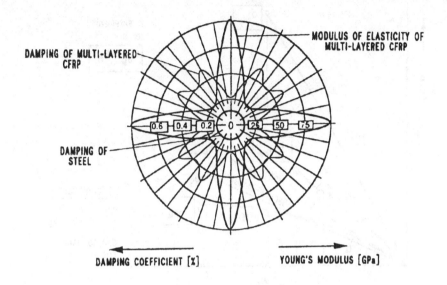

Figure 2.23. Polar diagram of stiffness and damping of multi-layered CFRP.

2.4 DEFORMATION LIMITS OF FRP

The good characteristics of the fibres and polymeric matrices are not always to be found in composites. Low or high fibre contents, brittle or heavily cross-linked polymeric matrices often lead to products which tend to generate early microcracks and therefore do not correspond to the required technical and physical demands. Mechanical compatibility is therefore an especially important criteria in a composite.

Mechanical compatibility means that fibres and matrix are homogeneously connected and have a common deformation under load, without any cracking. The stress-strain diagram of carbon fibres, polymeric matrices and fibre reinforced composites show deformation characteristics such as illustrated in figure 2.24.

The stress-strain behaviour of carbon and glass fibres are always linear up to the breaking elongation. The stress-strain curve of the matrix is not linear (viscoelastic) and its breaking elongation is normally higher than that of the fibres. Consequently, it could be assumed that the failure-free load of the composite extends to the fracture elongation of the fibres. Unfortunately, however, the first cracks are observed much earlier because of stress concentration. This

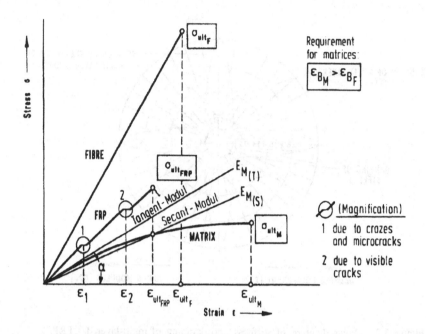

Figure 2.24. Stress-strain behaviour of fibre reinforced polymers.

has also been pointed out in [5]. Very early micro-cracks are generated, which are not visible with the naked eye. These are perpendicular to the load direction and grow quite fast to become cracks which are visible. These cracks do not alter the tensile strength very much, but nevertheless are irreversible damage. The elastic limits - where cracks first occur - are described by some authors as critical or allowable elongations and are an important factor in the choice of the structure. For high risk applications, such as for example in airplane construction, the elastic limit is set even lower by consideration of damage tolerance. An example is the safety regulation for structures for airplane construction according to [6] to be seen in figure 2.25, in which the allowable elongation limit is set as a function of the maximum damage probability.

For safety reasons it has to be ensured that in the airplane construction special structures fulfil their function even if partly damaged. For this reason the allowable elongations for high risk applications have to be reduced compared to less critical applications inside the plane. The diagram shows that for structures where damage is not likely to occur during the component lifetime an elongation

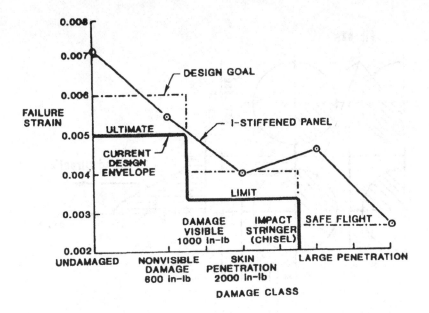

Figure 2.25. Damage tolerance characteristics, compression after impact.

strain of up to 0.5% is allowed. If higher damage is possible, for example at the leading edge of the wing, the allowable elongation strain is reduced to 0.25%. The main problem is tolerance to compression after impact, which sets the criterion for the damage tolerance of the material. Nevertheless there is a wide range of choice given to the design engineer. The ideal material would allow 0.6% elongation, but this is not yet possible due to a lack of damage tolerance.

Young engineers are surprised that the high potential strength of the composite is not used, because both the fibres and the polymeric matrices always show breaking-elongation of over 1%. The reasons are the low ductility and crack sensitivity as well as the lack of plastic deformation of the fibre reinforced thermosetting matrices. Breaks arise first between the matrix and the fibre reinforcement and take their course diagonally to the load direction. They occur because of stress concentrations on load transfer between fibres and matrix. The stress concentration-factor for glass fibres is at least four and therefore the matrix should have a four times larger ductility than glass fibres. The stress concentration-factor of the carbon fibres is less than four. A schematic is seen in figure 2.26.

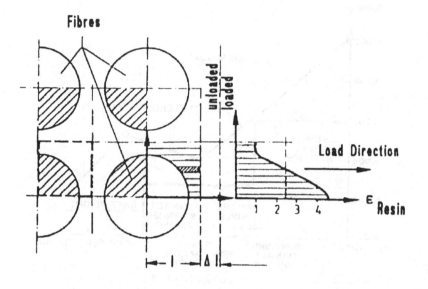

Figure 2.26. Micromechanical stress concentration during transverse tensile load.

2.5 OPTIMISING OF FIBRES AND RESINS

As mentioned in the previous section, cracks always occur at boundary surfaces and jump to the next stress concentration. Taking the cracking limit as a characteristic deformation limit there is, for each carbon fibre type, a special minimum-matrix-elongation, which is necessary to utilise the potential strength of the fibres. This is illustrated in figure 2.27.

Three different curves were measured for three different fibres. All composites with strong fibres show a large increase of the characteristic cracking limit with increasing ductility of the matrix. Specially strong fibres require even more than the 5% matrix-elongation of the presently available thermosetting matrices. But commonly used fibres - for example the T 300 (or T 3,6) - perform best with 3% resin-elongation. Beyond this there is no further significant gain. When using a very high modulus fibre - for example M 50 (or M 500) - no minimum desirable ductility of the matrix has been found. Using fibres with small elongation limits all the epoxy-resins tested have been found acceptable.

The results can therefore be summarised as follows:

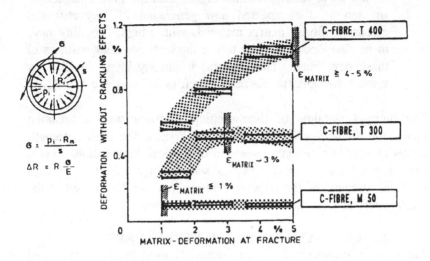

Figure 2.27. Characteristic deformation of CFRP without cracking (0% moisture).

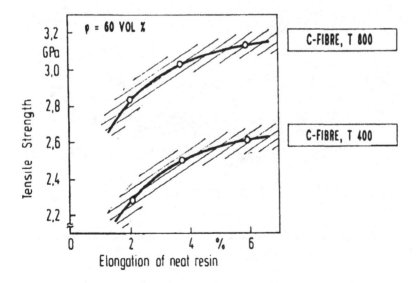

Figure 2.28. Effect of fracture elongation of matrix resin on tensile strength.

1. In a CFRP-laminate the fracture elongation of the matrix should be at least two times higher than the fibre elongation.
2. In view of the expected new generation of very stiff and extensible fibres, matrix materials with a higher ductility have to be developed, in order to use the high potential stiffness of these new fibres. This demand is already being fulfilled by some thermoplastic matrices - such as for example PEEK.

The thermal stability of thermosetting matrices with a breaking elongation above 3-4% are presently not known. The tensile strength seems to vary in a similar manner to the critical deformation. This is demonstrated by Japanese tests as seen in figure 2.28. The high potential strength of the improved fibres may not be utilised if the matrix materials can not be improved in ductility.

2.6 CORE AND SANDWICH STRUCTURES

The typical arrangement of core materials used in composites and sandwich structures is shown in figure 2.29. There is no standard relationship between the thickness of the skins and the thickness of the core. However, in most practical cases, the core is at least three or four times thicker than the skins.

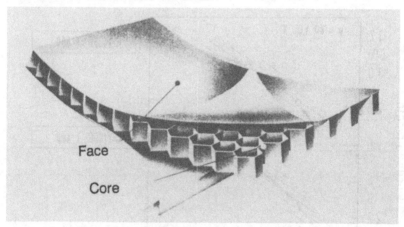

Figure 2.29. Typical sandwich structure with high strength skins and low density core.

The increase in both stiffness and strength, offered by the use of a composite core is easily illustrated by the variations in a simple sandwich shown in figure 2.30. For the purpose of illustration, a composite facing is chosen so that each skin has half the total thickness of a single laminate. The increasing bending stiffness and strength that can be achieved by the addition of a composite core placed in between the two half-thickness of the original laminate can be quite large, considering the small increase in weight involved. A brief outline of the different types of cores [7] is given below.

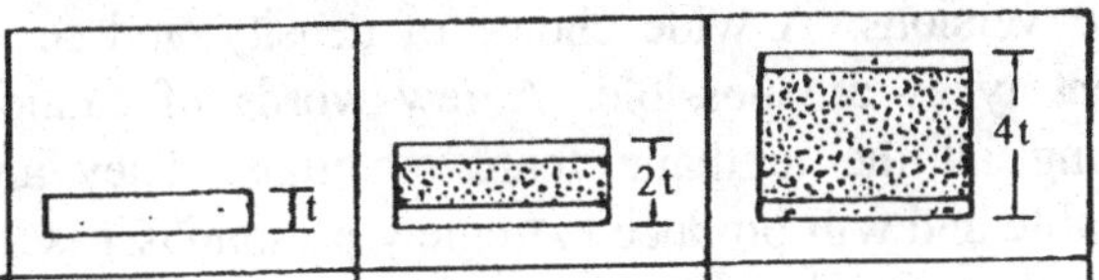

	t	$2t$	$4t$
Relative Stiffness (D)	1	7.0	37.0
Relative Strength	1	3.5	9.2
Relative Weight	1	1.03	1.06

Figure 2.30. Improvement in terms of stiffness and strength in sandwich structures.

Wood: Some of the most commonly used and least costly cores are light woods.The most popular is balsa. Substantial use has also been made of spruce, mahogany, redwood, pine, fir and many others.

The reason for the popularity of balsa is its good strength at densities as low as $128 \, kg/m^3$. When orientated in the sandwich so that the grain direction is perpendicular to the sandwich facings, the compressive strength of balsa, when measured at equal density, is higher than nearly all other cores, including some of the high performance honeycombs. Because of its excellent availability and reasonable cost, balsa has been used in more commercial, industrial and marine applications than all other wood cores combined.

Expanded Polymer Foams: Foamed plastic core materials are widely used in surfboards, boat hulls, as well as in commercial and amateur-built aircraft, because they are both reasonable in cost and easy to work with. Each member of this material family has a unique set of mechanical properties, physical properties, working and handling characteristics, as well as its own cost structure.

Polystyrene Foam: This is the material used in wings and for many homebuilt airplanes. Polystyrene foam is used because its strength properties are well matched to the needs of handling. It can be hot-wire contoured without generating poisonous gases, unlike the urethanes. It has the strong disadvantage of being soluble in styrene monomer, which is a major part of polyester and vinyl ester resins, and it is badly softened by exposure to gasoline and many other solvents.

Polyurethane Foam: This family of materials includes both rigid and flexible versions. A wide choice of density and several chemically different types are possible. A few words of caution are in order regarding all the urethane foam materials. They are usually very flammable and will produce extremely poisonous gases when burning. For this reason it is quite inadvisable to use hot-wire contouring for generating shapes. In addition these materials should never be incinerated or burned with other rubbish, even in an open area.

Polyvinyl Chloride (PVC) Foams: PVC foams are based on the same chemical family as the familiar garbage bags, plastic pipes and plastic film in common use, but are produced quite differently .The manufacturing process is more difficult and therefore the material is more expensive.

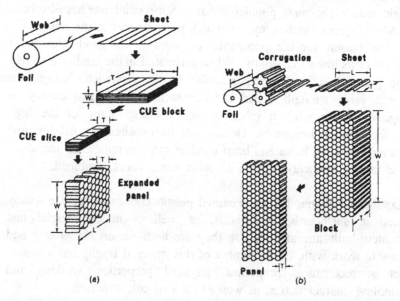

Figure 2.31. Two typical methods of making honeycomb; (a) expansion process and (b) corrugation process.

Honeycomb Core: These are made from impregnated paper or aluminium sheets. The two usual methods of manufacture, which are shown in figure 2.31, allow the finished core material to be furnished in large pieces. Commonly used cell shapes are illustrated in figure 2.32.

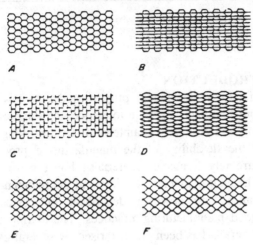

Figure 2.32. Most commonly used cell shapes in current honeycomb products.

2.7 REFERENCES

1] Niederstadt, G., "Die Viefalt der faserverstärten Polymere", Z. Werkstoff und Innovationen, 9-10/1990, 11-12/1990 und 1/1991.

2] Niederstadt, G., u. Mitautoren, "Leichtbau mit kohlenstoffaserverstärken Kunstsoffen", Expert-Verlag, Sindelfingen, 1985. pp 1-17.

3] Heissler, H. u. Mitautoren, "Verstärkte Kunstoffe in der Luft- und Raumfahrt", Kohlhammer-Verlag, Stuttgart, 1986. pp 172-198.

4] Niederstadt, G. u. Mitautoren, "Viskoelastizität und Dämpfung von CFK-Prepreg-Verbunden", Z. Kunstoffe, Hanser-Verlag, München, 80 (1), 1990. pp 65-69.

5] Puck, A., "Zur Beanspruchung und Verformung von GFK Mehrschichtenverbund Bauelementen", Z. Kenstoffe, Hanser-Verlag, München, 57 (12), 1967. pp 965-973.

6] Bohon, H.L., "Composites in Today's and Tomorrow's U.S. Airliners", Proc. Airmec Conf. Düsseldorf, 26th February - lst March 1985.

7] Lee, S.M. (ed.), "International Encylopaedia of Composites", VCH Publishers Inc., New York, 1990. pp 488-507.

3 ENVIRONMENTAL ASPECTS

3.1 INTRODUCTION

Composites offer high mechanical properties due to low density, a proven corrosion resistance, damage tolerance and fatigue resistance, leading to a long and relatively maintenance-free working life. This, combined with the flexibility of the manufacturing process, makes composites extremely suitable materials for use in a marine environment. The environmental stability and the maintenance of the initial performances are however dependent on correct material selection and good manufacturing practices. The marine application of composite materials has been characterised by several examples of failures and poor durability, such as osmosis, due to inadequate materials and working practices. The fire performance of composites is another topic which needs particular attention. It is well-known that composites offer the same if not a better level of protection to fire than traditional structural materials notwithstanding the fact that composite structures are often limited if not forbidden by current marine regulations for their combustibility and the consequent smoke emissions. Another important aspect is the safety and the health of people involved in boat and ship construction with these materials. In fact the chemical nature of these materials requires particular care both in the selection of materials and organisation of work. The scope of this Chapter is to give a review of the basic principles and the parameters which govern the behaviour of composite structures in a marine environment in order to provide the basic elements for a correct selection of materials, design and production techniques.

3.2 DURABILITY IN A MARINE ENVIRONMENT

The performances of composite structures under prolonged immersion and exposure to wind, rain and sun are generally good when compared with other usual construction materials. Despite such characteristics these materials suffer in some way a reduction of the original properties that, if not well understood and prevented can, lead to unexpected structural failures. These aspects should be carefully considered during the design process for a marine structure.

3.2.1 Water Absorption

Water absorption causes two macroscopic effects which lead to deterioration in the performances of a laminate namely, a general reduction of mechanical properties and a weight increase of the structure. The penetration of water in a FRP laminate occurs both by diffusion through the resin and by capillary flow through cracks and voids and along imperfect fibre-resin interfaces.

Degradation of mechanical properties caused by water absorption is attributed to plasticisation and consequent loss of stiffness in the resin, to debonding stresses across the fibre-resin interface induced by resin swelling and osmotic pressure and to chemical attack by the water on the fibre-resin bond. Experimental investigations carried out on specimens cut from laminates exposed to in-service conditions revealed that these effects can cause losses of strength and stiffness of up to 20%, most of which occur during the first few months of immersion. This consideration has suggested in the past the application of a partial safety factor of 1.1-1.2 to account for a long-term water-induced degradation of strength and modulus.

As a general approach for structural purposes, the long-term degradation of mechanical properties when composites are immersed in water can be found by applying curve-fitting programs to experimental data. Figure 3.1 [1] depicts a 25 year prediction of tensile and shear strength for glass polyester specimen dried after immersion.

The experimental assessment of the water absorption in composite materials is performed by measuring the water mass uptake M_t after a certain period; this measure is generally carried out using a thermogravimetric method. The mathematical model adopted to describe the process is the Fickian law:

$$\frac{M_t}{M_\infty} = 1 - \frac{8}{\pi^2} \sum_{\infty}^{n=0} \frac{1}{(2n+1)^2} \exp\left[\frac{-(2n+1)^2 \pi^2 D_t}{2h^2}\right]. \qquad (3.1)$$

Equation (3.1) assumes that diffusion into the laminate proceeds from surfaces which are in equilibrium with their surroundings. In this equation, D_t is the diffusion coefficient, M_∞ is the water taken at equilibrium and h is the thickness.

Obtained data are usually presented in the form of plots of fractional uptake (M_t/M_∞) or (M%) against $t^{\frac{1}{2}}$ or ($t^{\frac{1}{2}}/h$) for a certain

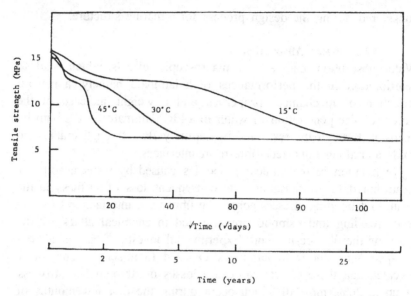

Figure 3.1. Predicted wet tensile strength versus square root of immersion time.

liquid and a certain temperature (T) which are described as absorption plots. A typical absorption plot is represented in figure 3.2 which shows the following features of Fickian absorption: the first part of the curve is linear and in this part D can be evaluated from the slope

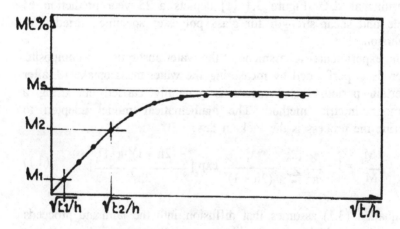

Figure 3.2. Typical water absorption curve.

using equation 3.2 [2]; above the linear portion the curve is concave against the abscissa and tends to M_∞ in an asymptotic way.

$$\text{Slope} = \frac{M_2 - M_1}{\sqrt{t_2/h} - \sqrt{t_1/h}} = \frac{4\,M_s\,\sqrt{D}}{\sqrt{\pi}}. \qquad (3.2)$$

In the following some of the parameters which affect the phenomenon are discussed:

- resin - water absorption in different resins follows the rule: phenolic > epoxy-based > vinyl ester > orthophthalic unsaturated polyester > isophthalic unsaturated polyester. Polyester undamaged laminates reach saturation at 20°C when water absorption is 0.5-1.0% by weight. Different considerations should be made on the diffusion rate and the strength retention. In fact vinyl ester and epoxy resins have a better behaviour (but also higher costs) than polyester resins while phenolic resins show a dramatic loss of mechanical properties when water aged. Several investigations have been carried out in the past in order to better understand the water absorption and the loss of strength. The results reported in current literature in general refer to the weight gain M due to the water immersion, to the diffusivity of the moisture in the laminate and/or the resin and to the residual mechanical properties after the exposure to the water. A comparative experience [3] reports the behaviour of two different polyester based and two epoxy-based laminates; the measurement of the residual flexural strength after 30,000 hours salt water immersion is about 75-80% (figure 3.3). A parametric investigation [4] carried out on four different polyester resins, see table 3.1, gives different values for D and M obtained at different temperatures and showed that isophthalic resins absorb less water than orthophthalic and tetrahydrophtalic resins. By adding thixotropic agents to isophthalic resins M is increased but D is decreased. The same investigation also gives the strength retention after fresh water ageing of six months at 60°C carried out by three-point bending tests (figure 3.4). The deterioration of mechanical properties follows the rule: ISO > TETRA > ISO-THIXO > ORTHO. Another parameter which has some influence on the absorption phenomena is the presence of fillers in the resin. A comparative study [5] between two laminates differing through the presence of a calcium carbonate filler shows how the presence of the filler can lead to a weight loss due to matrix dissolution and a consequent decrease in weight gain due to water absorption.

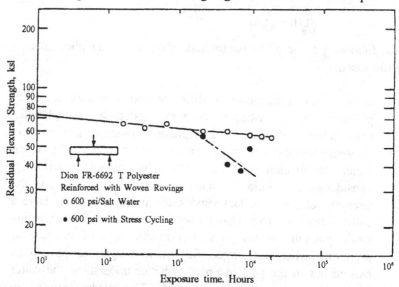

Figure 3.3. Flexural strength versus exposure duration for dion FR-6692T polyester reinforced by means of woven rovings and fabricated using wet layup processes.

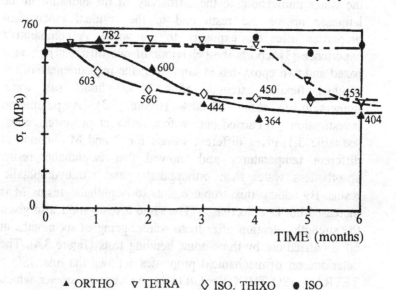

Figure 3.4. Influence of water ageing on ultimate strength of four different laminates (distilled water 60° C).

Table 3.1. Evolution of M_s and D for different temperatures (5,15,25,32,40°C).

MATERIAL	TYPE	5°C		15°C		25°C		32°C		40°C	
		M_s %	$D(mm^2/H)$	M_s %	$D(mm^2/H)$	M_s %	$D(mm^2/H)$	M_s %	$D(mm^2/H)$	M_s %	$D(mm^2/H)$
Polyester Orthophthalic	Pure Resin	1.22	$0{,}44.10^{-3}$	-	-	1.38	$1{,}07.10^{-3}$	-	-	1.4	$5{,}20.10^{-3}$
	In Laminates	0.586	$0{,}11.10^{-3}$	0.58	$0{,}18.10^{-3}$	0.61	$0{,}53.10^{-3}$	0.65	$0{,}93.10^{-3}$	0.66	$1{,}30.10^{-3}$
Polyester Tetrahydrophthalic	Pure Resin	1.25	$0{,}60.10^{-3}$	-	-	1.3	$2{,}02.10^{-3}$	-	-	1.44	$6{,}50.10^{-3}$
	In Laminates	0.5	$0{,}16.10^{-3}$	0.61	$0{,}31.10^{-3}$	0.61	$1{,}01.10^{-3}$	0.665	$1{,}58.10^{-3}$	0.62	$1{,}97.10^{-3}$
Polyester Isophthalic	Pure Resin	0.72	$1{,}20.10^{-3}$	-	-	0.75	$4{,}03.10^{-3}$	-	-	0.83	9.10^{-3}
	In Laminates	0.28	$0{,}47.10^{-3}$	0.35	$0{,}77.10^{-3}$	0.35	$1{,}07.10^{-3}$	0.37	$2{,}06.10^{-3}$	0.38	$3{,}40.10^{-3}$
Polyester Isophthalic-Thixotropic	Pure Resin	0.77	$1{,}20.10^{-3}$	-	-	0.81	$3{,}07.10^{-3}$	-	-	0.92	$7{,}80.10^{-3}$
	In Laminates	0.36	$0{,}40.10^{-3}$	0.38	$0{,}70.10^{-3}$	0.4	$1{,}06.18^{-3}$	0.45	$1{,}09.10^{-3}$	0.43	$2{,}60.10^{-3}$

- void content - because of the basic mechanism of water intake this parameter has great influence on the phenomena: the higher the void content level, the higher the water absorption. The void content is a consequence of the processing accuracy and the fibre percentage. The graphs from the Gibbs and Cox Manual [6] give the relationships between fibreglass content, voids and wet strength retention for a certain laminate (figure 3.5);

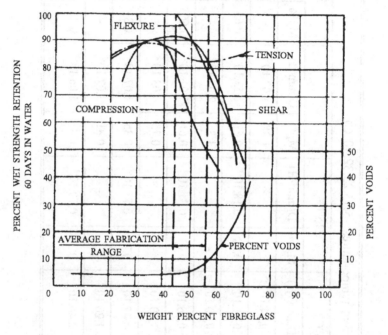

Figure 3.5. Fibreglass polyester laminates - contact molded 25-27 Oz. woven roving with silane finish. Relationships between fibreglass content, per cent voids and wet strength retention.

- water temperature - the relationship between diffusivity D and temperature is governed by the Arrhenius law:

$$D(T) = D_o \exp(-EA/RT) \tag{3.3}$$

D   = diffusion coeff. [mm²/hour]
EA = diffusion activation energy [°KJ/mol]
R   = perfect gas constant [J/mol °K]
D_o = constant function of the material and kind of the ageing
          liquid

T = temperature [°K].

Experimental investigations [4,7] performed on polyester laminates aged in fresh water at different temperatures from (5°-8°C) revealed an influence of the temperature on the weight gain (figure 3.6). For temperatures in excess of 40°C, absorption does not follow Fickian law. Tests performed on E-glass reinforced bisphenol based polyester resin subjected to 3 point bending test after a 1000 hour one side water immersion with variable temperatures (60-80-95°C) show that this parameter has a great influence on the change of the bending properties with immersion time (figure 3.7);

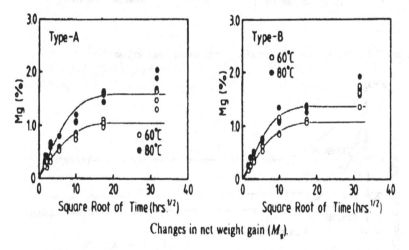

Changes in net weight gain (M_g).

Figure 3.6. Changes in net weight gain (M_g).

- pressure - water pressure influences the absorption rate but does not affect the minimum levels of strength retention. In fact pressure is used in accelerated long-term experiences as a satisfactory method. Data from reference [8] shows that a pressure of about 6.9 N/mm² (figure 3.8) leads to a quicker loss of mechanical properties.
- grade of protection - the previous example referred to samples which were not protected either in the surfaces or in the edges. Similar tests on sealed-edge samples revealed, after 14 weeks of immersion, that the laminates did not suffer of any loss of strength and modulus despite the presence of a water absorption level (recorded by weight increase) comparable with

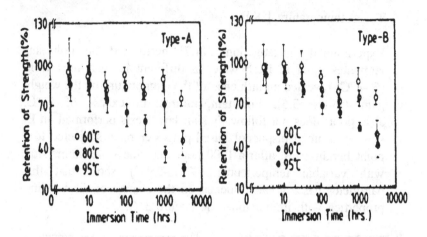

Figure 3.7. Relation between bending strength and immersion time.

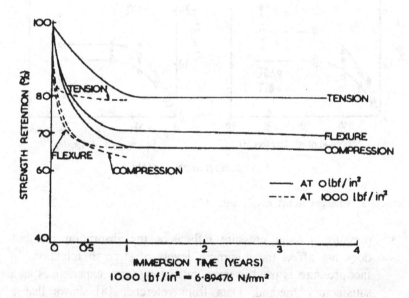

Figure 3.8. Strength retention of GRP immersed in tap water.

quantities recorded in the previous tests. This difference in
behaviour is probably due to the absorption of water affecting
the glass-resin bond rather than the resin itself and protection
of edges retarding the degradation of strength rather than
preventing it completely. It should be noted that most of the
reported experiences are referred to unprotected resins and

laminates; marine structures are always protected by means of a gel-coat and hardly exposed to the water on both surfaces. Water barrier capability of protecting coatings (gel coats) will be discussed in the section related to blistering.

- salt water/fresh water - a parametric study [4] performed on four different kinds of polyester laminates after the immersion in salt and fresh water at 40°C shows a lower saturation level and a higher diffusion coefficient in salt water, see table 3.2.

Table 3.2. Comparison of water absorption in sea water and distilled water at 40°C.

Material	Fresh Water			Salt Water		
	Slope mm/√h	D 10^{-3} mm²/H	M_s %	Slope mm/√h	D 10^{-3} mm²/H	M_s %
Orthophthalic	0.0539	1.3	0.66	0.0526	1.72	0.56
Tetrahydrop- hthalic	0.0618	1.97	0.61	0.0617	2.65	0.53
Isophthalic	0.0499	3.4	0.38	0.0519	3.8	0.37
Isophthalic- Thixotropic	0.0499	2.6	0.43	0.0478	3.6	0.35

There is some reported evidence of the long-term degradation of strength properties due to in-service conditions. In the early 50s US Coast Guard purchased a series of 40 feet single skin GRP Patrol Boat. In 1962 after 10 years of service in an extremely polluted channel (constant contact with sulphuric acid) some panels cut from three boats were statically tested. In 1972 a similar task was performed on the laminates cut from a boat retired after 20 years of service (11654 operating hours), during which the hull was exposed to extremely high temperature in a fire fighting episode. Table 3.3 [9] reports the obtained data. The comparison of mechanical properties after 10 and 20 years of service clearly depicts the long-term behaviour of the laminates employed. During the same period the US Navy developed a fibreglass fairwater of the USS Halfbeak submarine [10]; the motivation of the application was the electrolytic corrosion and maintenance problems recorded in the original aluminium solution. Laminates were made of glass reinforced polyester resin (blended with 10% of flexible resin to add toughness) with vacuum bag moulding (to assure low void content) and room temperature

Table 3.3. Physical property data for 10 and 20 year tests of USCG patrol boat [9].			
Hull CG 40503		10 Yr Tests	20 Yr Tests
Tensile Strength	Average psi Number of samples	5990 1	6140 10
Compressive Strength	Average psi Number of samples	12200 2	12210 10
Flexural Strength	Average psi Number of samples	9410 1	10850 10
Shear Strength	Average psi Number of samples	6560 3	6146 10

curing. Particular attention was paid to the choice of the finish in order to obtain high water resistance. Laminates were developed unpigmented in order to allow a visual inspection and void control. Tests were conducted on samples during the construction, after five years of service and after 11 years in service. Table 3.4 represents the results that show how the minimum required mechanical properties were maintained during all the service life. It should be noted that a detailed analysis of the component indicated that a safety factor of four was maintained throughout the service life of the part. Thus the mean stress was kept below the long-term fatigue limit, which at the time was taken to be 20 to 25% of the ultimate strength. This fact probably avoided stress corrosion effects.

3.2.2 Stress corrosion/Fatigue

A more serious form of degradation can occur in FRP subjected to immersion combined with continuous tensile stress. Stress corrosion of glass fibres appears to be caused by chemical action (attack by hydrogen ions), resulting in growth of microscopic cracks in individual glass fibres leading rapidly to fracture of the laminate at loads less than the 20% of the initial dry ultimate tensile stress. The fracture surface has a brittle appearance.

Table 3.4. Property tests of samples from fairwater of USS Halfbeak [10].

Property	Condition	Original Data (1954)*	1st Panel	1965 Data 2nd Panel	Average
Flexural Strength psi	Dry	52400	51900	51900	51900
	Wet †	54300	46400	47300	46900
Flexural Modulus psi x 10^{-6}	Dry	2.54	2.62	2.41	2.52
	Wet	2.49	2.45	2.28	2.37
Compressive Strength, psi	Dry	-	40200	38000	39100
	Wet	-	35900	35200	35600
Barcol Hardness	Dry	55	53	50	52
Specific Gravity	Dry	1.68	1.69	1.66	1.68
Resin Content	Dry	47.6%	47.4%	48.2%	47.8%

* Average of three panels
† Specimen boiled for two hours, then cooled at room temperature for one hour prior to testing.

In general, ship and boat structures are not greatly affected by this phenomenon since peak stresses induced by wave loading, slamming, etc., are of intermittent and short duration. This problem can affect local structures subjected to dead loads (bottom shell in the machinery area and tank bulkheads). In such cases the structure should be designed locally to very low stress levels based on experimental curves such as those of figure 3.9. Alternatively chemically resistant fibres (i.e. E-CR glass, CEM-FIL) should be selected.

Bisphenol polyester, vinyl ester and epoxy matrices which have lower water diffusion rate than polyester, delay but do not eliminate this problem.

Carbon fibres appear to be immune from stress corrosion effects while aramid fibres have an intermediate behaviour between that of E-glass and Cemfil.

E-glass/phenolic resin composites, because of the acid catalyst employed in the curing process, are suspected to be particularly susceptible to stress corrosion. Cold setting phenolic with chopped-strand mat and unidirectional E-glass reinforcement seems to have a behaviour equivalent to that of E-glass polyester laminates. The result of these tests was most probably influenced by the choice of the active agent in the catalyst. In fact, orthophosphoric acid employed in the reaction showed a degrading influence equivalent to that of water, in contrast with the rapid deterioration due to the use of the sulphuric acid.

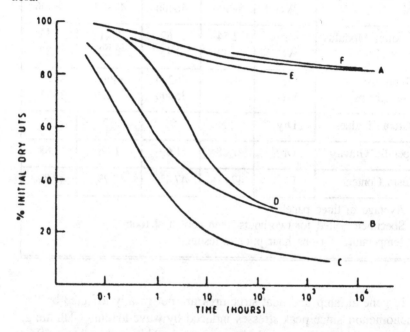

Figure 3.9. Stress-rupture of various GRP composites showing corrosion effects: A, E-glass/orthophthalic polyester, tensile load, in air - B, E-glass/orthophthalic polyester, tensile load, in water - C, E-glass/orthophthalic polyester, tensile load, in H_2SO_4 - D, E-glass/epoxy, tensile load, in water - E, CEM-FIL/orthophthalic polyester, tensile load, in H_2SO_4 - F, E-CR glass/isophthalic polyester, flexural load in water.

The exposure of composite structures to the marine environment induces detrimental effects on the fatigue behaviour too. The combination of sea water strength loss and stress cycling was investigated by aircraft industries. It was found that additional partial safety factors to the fatigue ones were necessary (a safety factor of 1.5 is usually assumed) [11].

Several experiments were performed in order to assess, by means of accelerated tests, the long-term behaviour of typical marine laminates. A comparison among two typical wet lay-up polyester systems (differing in void content, 3.5% for the first one and 0.5% for the second) and one rubber-toughened epoxy system was carried out [3]. The experiment was based on a short life (in cycles) of about 20,000 hours maintaining the actual temperature, environment and stresses in order to check current theory. Fatigue S-N curves were generated in tension, compression and in bending/shear stress mode in both dry and wet conditions. Figure 3.10 shows some of the results

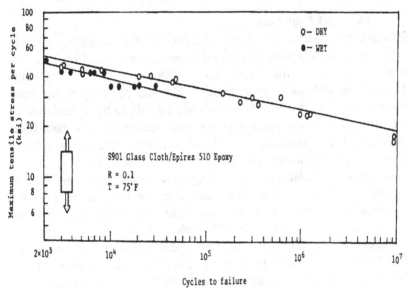

Figure 3.10. Effect of moisture on the S-N diagram for S901 glass cloth/epirez 510 epoxy laminate tested at P = 0.1, T = 75°F.

obtained in this context. As can be noted there is a progressive loss of strength for wet laminates with respect to the dry ones. Referring to typical laminates used in marine structures, major considerations can be summarised as below:

- current literature suggests partial safety factors (4-6) need to be applied in the design of long-term cyclic loaded structures for dry conditions. While in the aeronautical field the problem of cyclic loading in wet conditions was investigated, in the marine field there is a general lack of knowledge;
- for a refined design of structures exposed to sea water and

cyclic loads, experimental investigations are generally required
in order to avoid heavy partial safety factors;

- the higher the quality of the laminate, i.e. high volume-low
  void epoxy-based laminates, the easier it is to establish a
  definition of the long-term behaviour via experimental
  degradation curves and the adoption of moderate safety factors;
- the poorer the quality of the laminate, i.e. low density-high
  void boat construction polyester laminates, the more probable
  is an early loss of strength which calls for use of higher safety
  factors and more extensive testing.

3.2.3 Gel Coat Blistering

Although it is mainly a cosmetic problem, surface blistering of FRP
laminates, generally named "osmosis", has caused several problems to
owners and shipyards. The bad publicity derived from this
phenomenon influenced in a sensitive way the economy of the boat
construction field. More attention should be paid to the consideration
that osmosis in some cases indicates poor quality of both laminates
and production methodology which can lead to more serious problems
(such as high rate water absorption, stress corrosion, etc.).

Blisters occur in laminates after long-term, uninterrupted immersion
in water with a variable timescale from months to years. The
incidence of osmosis in boat construction has been indicated by
several studies as between 1-3%. The phenomenon had its maximum
incidence in the early years of GRP construction. For example, in
1973 5-10% of surveyed boats in the UK were found to show minor
blistering, while 1-2% required the total removing of the gel coat [12].
In Holland, in the period between 1973 and 1981, the incidence was
dramatically higher, at 22.5%.

Blisters occur immediately behind the gel coat or behind the initial
layer of glass as a result of air voids at interfaces within the laminate.
There are two main causes of blister development. The first involves
various defects produced during fabrication. Air pockets, mainly
localised in the gel coat, can cause blisters when a part is heated
under environmental conditions. Entrapped liquids are also a source
of blister formation. Table 3.5 from reference [13] shows some
contaminating sources and associated blister discriminating features.
The most frequent cause of blister formation is water penetration. The
osmotic process allows water molecules to penetrate through the gel
coat. The absorbed water fills up the small cavities existing at the
interface between gel coat and the laminate. Then the water reacts

with certain water soluble and water degradable substances present in resins (uncross-linked components), gel coat catalyst, accelerators, binder agents employed on the glass reinforcements or in the glass itself, to form an acid water solution trapped in the laminate.

Table 3.5. Liquid contaminate sources during spray-up that can cause blistering [13].

Liquid	Common Source	Distinguishing Characteristics
Catalyst	Overspray, drips due to leaks of malfunctioning valves.	Usually when punctured, blister has a vinegar-like odour; the area around it, if in the laminate, is browner or burnt colour. If the part is less than 24 hrs old, wet starch iodine test paper will turn blue.
Water	Air lines, improperly stored material, perspiration.	No real odour when punctured; area around blister is whitish or milky.
Solvents	Leaky solvent flush system, overspray, carried by wet rollers.	Odour; area sometimes white in colour.
Oil	Compressor seals leaking.	Very little odour; fluid feels slick and will not evaporate.
Un-catalysed Resin	Malfunctioning gun or ran out of catalyst.	Styrene odour and sticky.

This solution makes external water, via osmosis, enter the laminate leading to a rise of the pressure inside the cavities and to a further quantity of acid solution which calls more water. Finally the pressure reaches a critical value at which the bubble bursts. The gel coat in that area (osmotic blisters can have diameters well over 10 mm) is completely removed rendering the laminate completely unprotected and with water able to affect external layers. Blister formation in GRP is not easily attributed to one single factor but to a combination of factors. Furthermore this effect cannot be eliminated but only reduced or delayed by a control of these factors.

In the following paragraphs, the principal parameters which influence this phenomenon are summarised and commented upon:

- Environmental condition and production - workshop humidity must be less than 80%. Low resin content (<40%) and bad fibre impregnation promote blistering. During the consolidation care should be taken to ensure no air or moisture is trapped in the laminate. Laminate lay-up on gel coat should be performed at the correct time. An early lay-up causes a thickness reduction due to the solution of gel coat in the resin. A delayed lay-up causes defects in the bonding and poor interface bonding accelerates blister formation. A content of MEKP as catalytic agent both in the resin and in the gel coat in excess of 2% promotes blistering. In general it is suggested that a post-curing of about 16 hours at 45°C is required. An inadequate post-curing time leads to the presence of water soluble substances in the laminate and a loss in the gel coat thickness. Polyvinyl alcohol (PVA) which is the most common release agent has proved to strongly promote blistering. In general, Rules and Regulations issued by Classification Societies give the workshop environmental conditions which should be followed in order to reach good quality standards of products and avoid problems such as blistering.
- Resin composition - the higher the chemical resistance and molecular weight (>1100) the lower the risk of blistering. Presence of water-soluble residuals, particularly free glycols (avoid resins containing free propylenic glycol >0.6), and high resin acidity promote blistering.
- Glass - PVA based bond agents are to be avoided. Their presence in the laminate promotes the production of acetic acid which is water soluble. The presence of resin rich surface tissue immediately behind the gel coat reduces the chances of blistering as a thick gel coat does.
- Gel coat - the permeability is directly related to the gel coat thickness. Generally a value of 500 μm is suggested (about 600-700 g/m^2). Gel coat permeability should be equivalent or lower than that of the laminate to prevent the build up of water in voids at the interface. Generally this is obtained using a gel coat based on the same resin as the laminate. Dark pigments (i.e. blue and black) in the gel coat are to be avoided. Isophthalic acid-neopentyl glycol based gel coats are more

blister resistant coatings than orthophthalic acid and isophthalic acid low unsaturation based ones.

The evaluation of the blister resistance in a laminate is performed by means of accelerated testing. These tests consist of the exposure of FRP samples to hot water for a certain time and, through visual inspection, to assess the severity of the blister formation according to the hierarchy showed in table 3.6. The choice of water temperature leads to the choice of the acceleration factor. For example, a factor of 5-6 is reached with a temperature of 40°C. Temperatures up to 65°C are suggested for higher factors but a limit of 50°C is indicated for systems which have not been post cured at elevated temperature because of their low heat deflection temperature (55-60°C).

Several investigations [14,15] have been conducted in order to assess the blistering behaviour of typical marine laminates. The results are generally presented in figures representing the rapidity and the

Table 3.6. Classification of blister severity.

Type 1	Fibre whitening, no blisters
Type 2	As 1, but with small blisters
Type 3	Blisters 1-2 mm diameter, some burst
Type 4	Blisters 3-4 mm diameter, some burst
Type 5	Blisters 4-5 mm diameter
Type 6	Start of formation of blisters 5 mm diameter
Type 7	Blisters 6-10 mm diameter
Type 8	Blisters 10 mm diameter

seriousness of the phenomenon. Table 3.7 is a typical example taken from Norwood [12]. A recent theoretical and experimental investigation [16] examined the structural degradation effects of blisters within hull laminates. A finite element model of the blister phenomenon was created by progressively removing material from the surface down to the layers, as shown in figure 3.11. Strain gauge measurements were made on boat hulls that exhibited severe blisters. The authors of this work affirm that field measurements were in good agreement with the theoretically determined values for strength and stiffness. Stiffness was relatively unchanged, while strength values degraded by 15-30%.

Table 3.7. Time to blister formation of glass-reinforced isophthalic acid based resin composites.

Mat Type	Gel Coat Type (clear unless stated otherwise)	Surface Tissue	Time to Blister Formation (weeks)	
			Types 2 and 3	Types 4 and 5
	High HDT Marine Isophthalic Acid Based Resin			
Emulsion-bonded CSM	None	None	16	50
	None	Yes	18	44
Standard powder-bonded CSM	None	None	8	52
	None	Yes	36	> 52 (type 3 4 at 52)
Powder-sized powder-bonded CSM	None	None	12	> 52 (type 3 at 52)
	None	Yes	> 52 (type 1 at 52)	
Woven roving emulsion-bonded CSM	Medium HDT isophthalic (white)	None	22	52
Woven roving powder-bonded CSM	Medium HDT isophthalic (white)	None	40	
	Medium HDT isophthalic (white)	Yes	> 52 (type 0 at 52)	> 52 (type 3 at 52)

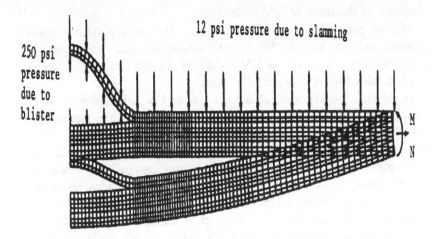

Figure 3.11. Internal blister axisymetric finite element model [16].

The blister resistance of different resins follows the law: orthophthalic > isophthalic > vinyl ester > epoxy. High cost, high performances resins, such as epoxy and vinyl ester, give tenuous benefits under the normal condition of cure and post-cure for marine laminates, offering little advantages over isophthalic acid based resins.

3.2.4 Weathering

Most GRP laminates employed in marine structures are protected by paint or a pigmented gel coat. Under these conditions, the effects of weathering action (of sun, wind, rain, hail and ice) are slight and generally affect only the cosmetic performances of the laminate leading to loss of gel coat gloss and superficial crazing, pin-hole blistering or erosion of the paint or gel coat. The loss of mechanical properties is generally less than that described in the previous paragraphs for moisture absorption or blister, etc.

Ultraviolet rays (UV) contained in the sunlight are considered to be the main cause of weather deterioration. The three major categories of resins that are used in boat building, namely polyester, vinyl ester and epoxy, have different reactions to exposure to sunlight. Epoxies are generally very sensitive to UV light and if exposed to UV rays for any significant period of time the resins will degrade to the point where they have little, if any, strength left to them. Vinyl esters are also sensitive to UV, as there are epoxy linkages in them and will degrade with time, although in general not as rapidly as epoxies. Polyesters, although they suffer some degradation, are the least sensitive of the three to UV light.

Marine grade gel coats are based on ortho or isopolyester resin systems that are heavily filled and contain pigments (for most gel coats the pigment serves as UV protector). In addition, a UV screen is often added to help protect the resin.

In general, the exposure of the gel coat to UV radiation will cause fading of the colour which is associated with the pigments themselves and their reaction to sunlight can sometimes cause yellowing. The yellowing, in general, is a degradation of the resin rather than the pigments and leads to a phenomenon known as chalking. Chalking occurs when the very thin outer coating of resin degrades under the UV light to the point where it exposes the filler and some of the pigments in the gel coat. The high gloss finish is due to that thin layer. Once it degrades, the gloss is gone and the surface is still coloured but not shiny. Because the pigments are no longer sealed by the outer layers of the resin they lose their colour and can loosen up from the finish to give a kind of a chalky surface effect.

In addition, sunlight exposure can lead to various effects due to the heat. Firstly the thermal expansion coefficient of fibreglass is very different from that of the resin; thus when a laminate with a high glass content is heated significantly the resin tries to expand but it is held in place by the glass. The result of this is that in many cases the pattern of the fibreglass will show through the gel coat a phenomenon known as print through.

Another effect of the heating laminates is related to the curing temperature and the heat distortion temperature. Most polyester resins have a heat distortion temperature of around 65-95°C, which means that when the resin becomes heated to that temperature it becomes very soft and consequently the laminate becomes unsuitable. In addition, the heated resin can cure further and when it cools will try to shrink, but the glass holds it in place creating very large internal stresses solely due to these thermal effects. This can be a problem for all room temperature thermosetting resins. As an example, in the tropics it is not uncommon to get temperatures in excess of 65°C on boats with white gel coats, while temperatures up to 85°C have been measured with red gel coats, and well over 95°C with black gel coats. Deterioration can be accelerated because during the day the laminate is heated and during the night is cooled down. These temperature cycles tend to produce internal stresses which then cause the laminate to fatigue more rapidly than it would normally.

Another thermal effect on fatigue is caused by shadows moving over the deck of a boat that is sitting in the sun. As the sun travels overhead, the shadow will progress across the deck. At the edge of the shadow there can be a very large temperature differential, of the order of 20-30°C. As the shadow line travels there is a very sharp heating or cooling at the edge and the differential causes significant stress at that point. These effects show how clear or white gel coats are preferred to dark or black ones in boat construction.

Cold can also represent a cause of troubles. Most resins absorb some amount of moisture; a laminate which has absorbed a significant amount of moisture can experience severe stresses when frozen, since water expands inside it. This expansion can actually cause delamination or stress cracking.

Apart this problem it should be said that FRP laminates show, in general, a very good behaviour at low temperatures. In fact, as can be seen in figure 3.12 [17] the mechanical properties of usual marine

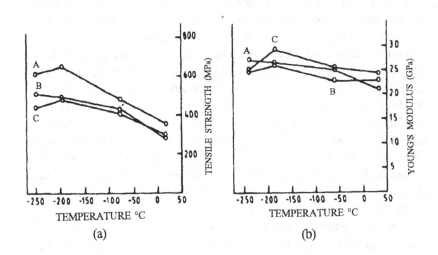

Figure 3.12. Strength (a) and stiffness (b) of GRP laminates at low temperature: A, glass cloth/epoxy; B, glass/cloth polyester; C, glass cloth/phenolic.

laminates are maintained, if not improved, for temperatures down to minus 250°C. This characteristic, which has proved valuable in aerospace application of composites, suggests further potential applications for structures subjected to Arctic conditions (offshore) and in cryogenic structures (LNG containment). Test data for carbon-epoxy laminates show reductions of 10-25% in tensile strength as temperature is reduced from 20°C to -200°C [18,19]. The tensile strength of aramid-epoxy laminates and the Young's modulus of both types of composite are virtually unchanged over the same temperature range.

3.3 CHEMICAL RESISTANCE/SAFETY AND HEALTH

3.3.1 Chemical Resistance

GRP pipes and tanks have been employed since early 60s in chemical and sewage plants. This successful in-service experience is evidence of the chemical resistance of these materials. Polyester based components, when properly cured, can be used without special protection for containment of salt and fresh water, fuel oils and gasoline. For these applications no deterioration or damages are reported in the current literature [6,17].

Some case studies are reported by the commercial literature [20]. In 1963 a large US oil company put in service a series of composite underground gasoline tanks as an alternative to steel ones which were prone to leakage and corrosion. These unsaturated isophthalic polyester based containers were unearthed after various periods of time from 5.5 up to 25 years; samples were cut from the laminates and subjected to a mechanical characterisation. The results obtained are displayed in table 3.8. During the development of HMS WILTON accelerated tests were performed in order to establish the capability of GRP tanks to resist to fuel and hydraulic oils [8]. GRP tanks were filled with diesel fuel at 60°C for 95 days, after which strips were removed for mechanical tests. No evidence of deterioration was observed; similar tests were carried out with two types of hydraulic fluids for 21 days with the same result.

Polyester resins are attacked by strong alkalis, by oxidising acids and by certain solvents which are present in cleaners, paint removers, primers and adhesives, including acetone and trichloroethylene [21]. These elements can cause the stress corrosion problems discussed in section 3.2.2.

In particular, it has been found that all types of paint removers have a significant effect on laminates [8]; trichloroethylene which is a common component of these elements, has been shown to reduce the flexural strength of GRP by 32% after a 24 hour immersion.

A comparison of the behaviour of different commercial polyester and vinyl ester resins, after one year's exposure to various chemical agents, is proposed by [20] and displayed in table 3.9.

Table 3.8. Properties comparison of unearthed tanks.*

		Tank A	Tank B	Tank C
Age at testing		5.5 years	7.5 years	25.0 years
Buried-excavated		1.7.65-8.21.70	4.4.64-10.24.71	5.15.63-5.11.88
Flexural strength,	psi	19,5000	24,200	22,400
	MPa	134	167	154
Flexural modulus,	psi	725×10^3	795×10^3	635×10^3
	MPa	4992	5482	4378
Tensile strength,	psi	10,700	13,600	10,500
	MPa	74	94	72
Tensile modulus,	psi	$1,160 \times 10^3$	$1,053 \times 10^3$	$1,107 \times 10^3$
	MPa	7260	8000	7630
Tensile elongation,	%	1.11	1.25	1.13
Notched Izod impact strength,	ft-lb/in	9.7	11.0	14.1
	J/m	518	587	753

* Results listed are averages of several samples. All tanks made from isopolyester resin SG-10 (described in Bulletin IP-86) with glass fibre reinforcement.

Table 3.9. Comparison of commercial resins after one year's exposure.

	Commercial Isopolyester		Atlac 382 Bisphenol A Polyester		Derakane 411-45 Vinyl ester		Atlac 4010 Bisphenol A Polyester		Heltron 197 Halogenated Polyester	
	Flexural Strength MPa	Flexural Modulus MPa	Flexural Strength MPa	Flexural Modulus MPa	Flexural Strength MPa	Flexural Modulus MPa	Flexural Strength MPa	Flexural Modulus MPa	Flexural Strength MPa	Flexural Modulus MPa
Initial	115	5800	95	5000	160	5800	110	4500	140	7100
Distilled Water	90	4600	75	4100	100	5000	70	4000	90	5300
5% Nitric Acid	75	4200	85	4500	95	4000	85	4100	80	5200
5% Hydrochloric Acid	92	4800	68	4200	115	5000	75	4100	75	5100
25% Sulfuric Acid	70	5400	70	4800	125	5300	80	4300	85	5500
Fuel Oil	98	5400	95	4800	140	4500	110	4500	95	6100
Ethyl Gasoline	90	5100	92	5000	145	5000	100	3000	105	6500
Unleaded Regular	110	5800	82	5000	110	4000	95	3500	95	6000
Benzene	90	3500	58	1800	40	900			50	1500

Note. The listed values are only indicative of the influence of different chemical agents.

The chemical resistance of epoxy and vinyl ester resins is generally superior to that of polyesters [22].

3.3.2 Safety and Health

Workers exposed to the typical environment of an FRP workshop can be prone to dermatitis and irritations to skin, eyes or respiratory passages due to the direct contact with the handled materials or their emissions during the moulding and the curing processes.

The emission of styrene vapour that occurs during the cure of polyester resins is one of the major problems related to the workers' health. Styrene, in large concentrations, has an irritant effect on the eyes and respiratory passages and a depressant effect on the nervous system. As will be discussed below national societies for control of the health establish threshold limits for styrene concentration. This limit leads to the need for a comprehensive ventilation system, in particular in the moulding area, together with means of monitoring styrene levels.

The use of proper ventilation is the primary technique for reducing airborne contaminants. There are three types of ventilation used in polyester fabrication shops:

- General (dilution) ventilation; the principle here is to dilute contaminated air with a volume of fresh air. These types of systems can be costly as the total volume of room air should be changed approximately every 2 to 12 minutes;
- Local ventilation; a local exhaust system may consist of a capture hood or exhaust bank designed to evacuate air from a specific area; spray booths are an example of local ventilation devices used in shops where small parts are fabricated;
- Directed flow ventilation; these systems direct air flow patterns over a part in relatively small volumes. The air flow is then captured by an exhaust bank located near the floor, which establishes a general top-to-bottom flow.

When prepregs (or pre-impregnated material) are employed the problem of toxic emission is generally overcome as the curing process is carried out in an oven and/or with vacuum bag techniques. In this way the workers are not exposed to the direct emission of the curing products.

Another precaution to be taken is to avoid the direct contact of unprotected areas (hands, lower arms and face) with the raw materials (fibres and resins) and other chemical agents employed in the process (diluents, accelerators, promoters, etc.). In particular, products contained in prepregs are very dangerous to the health. The proper type of gloves, shoes, glasses and clothes should be worn, barrier creams and resin removing creams should be provided during the process and proper medical aid for injured should be provided too. In addition to these precautions, is also essential to use face masks during finishing and cutting operations of cured laminates.

3.3.3 Regulatory and Statutory Aspects

In general, each nation has its regulations concerning the workers' health. Indications are given on the control and maintaining of the quality of the air and on the maximum concentration and time of exposure to toxic agents. For example, in the UK, the Factory Act stipulates a 100 ppm as the maximum concentration limit of styrene in the workshop. In the USA, limits are stated to control the workers exposure to hazard chemical agents; these exposure limits are based on standards developed by the American Conference of Governmental Industrial Hygienists (ACGIH) [23]:

- Threshold Limit Value - Time Weighted Average (TLV-TWA) - the time-weighted average for a normal 8 hour work day and a 40 hour work week, to which nearly all workers may be exposed, day after day, without adverse effect.
- Threshold Limit Value - Short Term Exposure Limit (TLV-STEL) - the concentration to which workers can be exposed continuously for a short period of time (15 minutes) without suffering from (1) irritation, (2) chronic or irreversible tissue damage, or (3) narcosis of sufficient degree to increase the likelihood of accidental injury, impair self-rescue or materially reduce work efficiency, provided that the daily TLV-TWA is not exceeded.
- Threshold Limit Value - Ceiling (TLV-C) - the concentration that should not be exceeded during any part of the working day.

The Occupational Safety and Health Administration (OSHA) issues legally binding Permissible Exposure Limits (PELs) for various compounds based on the above defined exposure limits. The limits are published in various reports [24,25]. In table 3.10, the permissible limits for some agents in a composite fabrication shop are reported.

Table 3.10. Permissible exposure limits and health hazards of some composite materials [23].

COMPONENT	PRIMARY HEALTH HAZARD	TLV-TWA	TLV-STEL
Styrene Monomer	Styrene vapours can cause eye and skin irritation. It can also cause systemic effects on the central nervous system.	50 ppm	100 ppm
Acetone	Overexposure to acetone by inhalation may cause irritation of mucous membranes, headache and nausea.	750 ppm	1000 ppm
Methyl Ethyl Keytone (MEK)	Eye, nose and throat irritation.	200 ppm	300 ppm
Polyurethane Resin	The isicyanates may strongly irritate the skin and the mucous membranes of the eyes and respiratory tract.	0.005 ppm	0.02 ppm
Carbon and Graphite Fibres	Handling of carbon & graphite fibres can cause mechanical abrasion & irritation.	10 mg/m^3 *	-
Fibreglass	Mechanical irritation of the eyes, nose and throat.	10 mg/m^3 †	-
Aramid Fibres	Minimal potential for irritation to skin.	5 fibrils/cm^3 ‡	-

* Value for total dust - natural graphite is to be controlled to 2.5 mg/m^3.
† Value for fibrous glass dust - although no standards exist for fibrous glass, a TWA of 15 mg/m^3 (total dust) and 5 mg/m^3 (respirable fraction) has been established for "particles not otherwise regulated".
‡ Acceptable exposure limit established by Du Pont based on internal studies.

3.4 FIRE AND TEMPERATURE RESISTANCE

3.4.1 Fire and People

Fire is an oxidation reaction at the surface of a material which is releasing reactive vapour. Three components are needed to permit burning to start, and once started to continue to burn. These are a combustible surface, an oxidising agent and a source of heat. The close relationship among these factors is conventionally illustrated as "the fire triangle" (figure 3.13). The most critical results of the

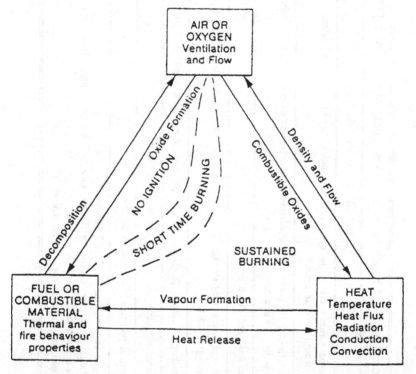

Figure 3.13. The fire triangle.

burning process onboard a ship are the threats to human life, wherever passengers and crew are present within the burning system and in adjacent compartments. However, their importance can actually vary with conditions of each individual fire situation. The main cause of threat to human life can be summarised as the following:

- oxygen depletion: the average human being is accustomed to

operating satisfactorily with the usual level of about 21%
oxygen in the atmosphere (see table 3.11).

Table 3.11. Response of humans to various oxygen concentrations.

Concentration %	Symptoms
21	Normal concentration in air.
17	Respiration volume increased, muscular coordination diminished, more effort required for attention and clear thinking.
12 to 15	Shortness of breath, headache, dizziness, quickened pulse, quick fatigue upon exertion, loss of muscular coordination for skilled movements.
10 to 14	Faulty judgement.
10 to 12	Nausea and vomiting, exertion impossible, paralysis of motion.
6 to 8	Collapse and unconciousness but rapid treatment can prevent death.
6 or below	Death in 6 to 8 minutes.
2 to 3	Death in 45 seconds.

- flame: burns can be caused by direct contact with flames or by heat radiated from flames; burns can result if skin temperature is held above 65°C for one second; flame temperature and their radiant heat may also prove to be fatal.
- heat: unlike the direct flame, heat can be a hazard to occupants of the burning system; a breathing level temperature of 148°C is considered to be the maximum value for survival.
- fire gases: while the toxicity of some gaseous products of combustion is well-known, the concentration of these gases in an actual fire is not so well-known, even from simulated conditions in laboratory experiments.
- smoke: the principal hazard of smoke is that it hides the escape of occupants and the entry of fire-fighters seeking to locate and extinguishing the fire;
- structural strength reduction: the failure of structural components through heat damage or burning can present a serious hazard; perhaps the most dramatic examples are the collapse of weakened decks under the weight of fire fighters, and the collapse of bulkheads on people beneath them.

A plastic is defined as a material which contains as an essential ingredient an organic substance of large molecular weight. Because of this, combustion can occur under sufficiently severe exposure to heat and oxygen (figure 3.14).

Furthermore these materials suffer damage even if only subjected to elevated temperatures and such damage can be hazardous to the safety and the health of both passengers and crew.

These problems need to be addressed adequately if the application of composite structures in the boat and shipbuilding industry is to be enhanced.

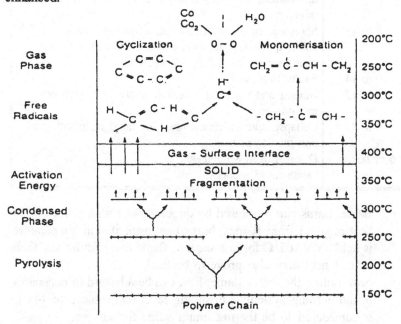

Figure 3.14. Reaction of plastic materials to heating.

3.4.2 Regulatory and Statutory Aspects

The application of composite structures in the marine industry, as with all the structural materials, is regulated by the norms issued by the International Maritime Organisation - IMO. The general requirements stated by this organisation can be summarised in the following four points:

- fire reaction: the capacity of a material in specified conditions to have or not a share in the fire phenomena particularly by its own alteration

- fire resistance: the ability of a certain structure to resist, for a defined minimum time, degradation of its intended purpose, in this case structural integrity
- containment: the ability of a material and the deriving structures to act as fire, temperature and smoke barrier, during a fixed period of time and under specified conditions
- smoke and toxicity: the production of smoke and toxic fumes when exposed to fire

Following the hierarchy stated by IMO it is possible to subdivide the analysis of regulatory aspects into three fields: namely large commercial ships, high speed vehicles and small commercial and pleasure vessels.

Large commercial ships: as far as large commercial ships (passengers and cargo ships having GRT > 500) are concerned, the in-force regulation is the International Convention for the Safety of Life at Sea [26]. This convention states that the hull and the superstructures shall be constructed in steel or other equivalent material, implying that any structural material shall be at least non-combustible. This requires that an unprotected sample of the structural material shall be exposed, in a oven, to a temperature of 750°C without burn and release inflammable vapours in such quantities as to get spontaneously ignited. Furthermore the mean weight loss measured after the test shall not exceed 50% of the initial weight.

Following the certification of this basic characteristic most of the ship structures need to comply with the other above mentioned requirements. In particular, primary and secondary structures which provide division for a certain number of areas and spaces, are certified as "A" and "B" class division by means of a standard fire test. This test consists of the exposure of a mock-up of the structure to fire, and is obtained by putting the sample at the opening of a special furnace in such a way that the sample is the closure and by making the temperature follow a time dependent curve which reaches the maximum value of 925°C after 60 minutes (see figure 3.15). Temperature measurement inside the furnace and in the external side of the sample are taken during the test.

"A" class divisions are defined as structures which are suitably stiffened, are capable of preventing the passage of flame and smoke to the end of the one-hour standard test (half-hour for the "B" class) and provide an insulation such that the average temperature of the unexposed side will not rise more than 139°C above the original

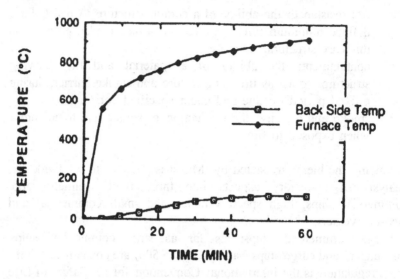

Figure 3.15. Temperature rise on unexposed face of a 30 ply - 25mm polyester woven roving laminate.

temperature, not in any point the rise more than 180°C (225°C for "B" class) above the original temperature. The period of time during which the division can comply with these requirements defines the grade such as A-60,30,15,0 (B-15,0), etc.

Composites usually employed in the marine industry are combustible and most of them produce large quantities of dense smoke. However, it is relatively simple to achieve A15,30 and even A60 ratings in respect to temperature rise and burn-through using composite structures even without mineral wool barrier materials (the use of these barriers reduces the thickness of the panel required to achieve the rating).

The direct experience of the author is that by working on composition and curing process it is possible to obtain laminates that can comply with the non-combustible SOLAS statement. However, this characteristic is obtained by reducing the mechanical properties so that the resulting laminates can be applied only in secondary structures subjected to local and limited loads (B and A0 classes). Further improvement in mechanical properties is possible but this development leads to very expensive structures.

For these reasons, although several studies for the development of large GRP structural components have been carried out [27,28,29], actual composites on board passengers and commercial ships are limited to the construction of secondary non-structural components or fittings (masts, funnels, internal and external surfaces, etc.) and for limited length of pipe systems. A similar situation is reported in the offshore field in which the known applications are for the realisation of pipe systems, gangways, and deck gratings.

The usage of composite components for pipe systems have been subject of particular attention of IMO which has published an information notice [30] about requirements for the usage of materials other than steel for the fabrication of piping. These requirements define three levels on the base of the fire resistance of pipes subjected to a specified fire incident (1098°C after 1/2 hour, 1100°C after 1 hour):

- Level 1 (LV1) - the pipe shall not show any loss of integrity in dry (empty) condition for 1 hour
- Level 2 (LV2) - half-hour having the same condition of LV1
- Level 3 (LV3) - half-hour in wet conditions (liquid in the pipe)

Recent investigations by the author show that it is possible to reach levels LV1 and LV2 but that resulting products are generally too expensive and of large dimensions to be applied onboard while LV3 is quite simple to reach with fairly economical products.

High speed vehicles: in 1977 IMO adopted the "Code of Safety for Dynamically Supported Craft" (DSC) [31]. This Code concerns high speed craft (with Froude number greater than 0.9) which carry less than 450 passengers, involved in voyage less than 100 nautical miles from a place of refuge, provided with special systems of safety and aid from both onboard and in-shore. For this class of vehicles the philosophy is to achieve an equivalent safety level to that of conventional displacement ships even though some requirements, such as those regarding the fire resistance of structures and lifeboats, are not satisfied. This normative approach has led to the construction of several FRP high speed vessels, in particular Surface Effect Ships, having lengths up to 50 m and speeds up to 55 kn.

From current reports it seems that structural fire protection needs to be ensured using approaches similar to those of SOLAS. This means that non-combustible materials are required for the construction of main hull and superstructures unless national authorities accept exemptions for the use of combustible materials. If such an approach is followed in any new code, the structural application of FRP in the high speed vehicles will be strongly restricted.

Table 3.12. Shipping rules - fire protection.

	Ships > 500 GRT	Ships < 500 GRT	Fishing ships L > 55 m	Fishing ships L > 24 m - L < 55 m	Dynamic Lift Vehicles	Yachts L < 25 m
Use of composite materials - bearing structures (Deck, hull, bulkhead)	NO	YES	NO	YES	YES	YES
		Passengers (1st or 2nd class) for cargo ships				
	Passengers and cargo ships making national trips which are not covered by SOLAS convention					
	International rules SOLAS convention (Booklet 1308) Safety of life at sea) IMO 74		International rules Torremolinos Convention	International rules Resolution A 373 IMO		
	Texts application Resolution A 472 IMO (inflammability) Resolution A 517 IMO (bulkheads)		Texts application Resolution A 517 IMO	Texts application Resolution A 517 IMO	Texts application Resolution A 517 IMO Resolution A 472 IMO	

Small commercial and pleasure vessels: for small commercial (fishing and passengers) and pleasure vessels, generally not exceeding 50 m in length, the use of FRP is permitted without significant restrictions. Indications are given by international [32] and national regulations about the minimum fire reaction capability of the structural materials and about meanings and systems to be adopted in order to reach the necessary fire safety level.

In table 3.12 the actual normative situation for the above mentioned vessel types is summarised.

3.4.3 Effects of High Temperature

The mechanical properties of both resins and fibres are affected quite differently by increase in temperature. Carbon fibres [17] retain most of their strength and stiffness at temperatures over 1000°C (provided that oxygen is excluded). The strength of E-glass, relative to its strength at 20°C, is reduced to about 75% at 350°C and 50% at 500°C. The strength of Kevlar 49, again related to that at 20°C, is 75% at 200°C and 60% at 300°C.

Organic resins suffer gradual thermal degradation in mechanical properties until they are exposed to temperatures between 350 and 500°C. After the self-ignition temperature is reached, the change becomes dramatic and rapid.

The tensile strengths of unidirectional composites loaded in the fibre direction have a temperature dependence similar to that of the reinforcing fibres. Under compressive, flexural, shear and transverse tensile loads, temperature dependence of mechanical properties is dominated by matrix behaviour.

Figure 3.16 shows the variations in Young's modulus and compressive strength with temperature in glass-fabric reinforced laminates employing polyester, vinyl ester and phenolic resins. As can be seen the performances follow the rule: phenolic > vinyl ester > polyester.

For higher temperatures, the degradation results are more critical. In fact all the resins systems suffer self-ignition and burn. In order to better understand the effects of the thermal insulation on laminates in this range of temperatures, it is useful to refer to a thermal-gravimetric analysis [27].

For example, referring to figure 3.17, the behaviour of a polyester resin (vinyl ester and epoxy resins have similar behaviours) under thermal insulation can be summarised as follows:

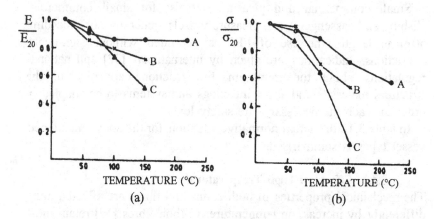

Figure 3.16. Effect of elevated temperature on stiffness (a) and strength (b) of GRP laminates with various matrix materials: A, phenolic; B, vinyl ester; C, polyester.

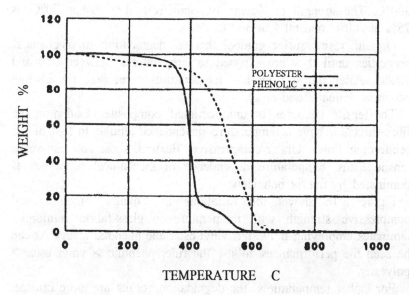

TEMPERATURE C

Figure 3.17. Ramped temperature thermal gravimetric analysis plots in air for a polyester and phenolic resin.

- insignificant degradation occurs up to 150°C
- in the region of 300°C-400°C the resin is rapidly degraded and loses the majority of its mass
- above 400°C in a reactive environment (air) the resin char is ultimately pyrolysed

The same analysis applied to a phenolic resin shows enhanced tolerance to thermal effects. Except for some initial mass loss due to water release, the phenolic resin will lose less mass than the polyester one for a given temperature and time of exposure. For example at 400°C the polyester loses 60% of its mass whereas the phenolic only 10%! Furthermore phenolic resins have a higher self ignition temperature (about 570°C) than that of other resins (about 480°C for polyester).

3.4.4 Fire Behaviour
It has already been said that the fire performance properties of composite materials are the major concern limiting their use in ship construction, although there is evidence of the global good fire behaviour of such a structures.

During a fire accident which occurred in the engine room of HMS Ledbury (a Hunt-Class MCMV) some facts were noted:

- a significant diesel oil fire in a machinery space occurred
- high temperatures are thought to have been achieved within the compartment - certainly well in excess of 600°C
- apart from some burn-through via a gland into an adjacent compartment, the low thermal conductivity of the GRP prevented the fire spreading to adjacent compartments
- adequate cooling of structure was difficult and re-ignition of the structure occurred in several instances
- the repair of the damaged structures was not difficult

The fire lasted for some four hours before it was extinguished with temperatures sufficient to melt aluminium fittings and to cause severe charring of the laminate to a depth of several millimetres. The remaining thickness of shell and bulkhead laminate was found to have virtually unimpaired mechanical properties and paint on the reverse side was not even discoloured.

Several investigation and test programmes intended to evaluate and demonstrate the fire barrier and resistance capability of composite structures are reported in the current literature [27,33,34]. All these tests have been carried out both on simple panels and on large complex structures (full scale compartments). In general they have confirmed the behaviour shown in the above mentioned accident.

As a general result it has been found that sandwich panels offer particularly good thermal insulation and residual mechanical properties but some precautions should be taken in order to avoid dramatic failures such as:

- failure of the core-face bond due to excessive heating of the adhesive
- decomposition of the foam core material involving large expansive vapour emission
- rupture of the honeycomb cells due to thermal induced high pressures

Test results indicate that dramatic improvement in the fire resistance performances can be obtained by providing the structure with proper means of protection such as phenolic based layers, ceramic or rock-wool materials.

Presently available fire retardant agents such as chlorine, bromine or antimony-based additives offer to the resins reduced flammability but affect laminate mechanical properties and wet durability.

Another indication obtained is that there is an urgent need to establish test methods for the evaluation of the effective mechanical behaviour and the residual load bearing capability of structures subjected to fire. These methods, which have been recently introduced in the civil and railways (channel tunnel) fields, should consist of the combined fire incident and mechanical loading in order to simulate the real conditions. Unfortunately this experiment is very costly and difficult to carry out. On the other hand alternative non-combined methods in which the sample is subjected to fire and mechanical loads in different times have showed considerable limitations.

3.4.5 Smoke Emissions
Shipboard fires are among the most difficult fires to fight, primarily because of the amount of smoke. The dense smoke generated can reduce the visibility and cause confusion and disorientation if not be the cause of death (as in the case of the ferry Moby Prince burnt in the Tirrenian sea in April 1991).

Recently international and national authorities have begun to consider the problem of smoke emission in fire conditions and the related toxicity from a qualitative point of view. The adoption of standard tests has been suggested by several nations.

Polyester as well as vinyl ester and epoxy-based laminates burn slowly in air but with large emission of dense smokes and low oxygen index (25 to 35). Phenolic resins, due to their particular chemical nature, are characterised by a higher oxygen index (45 to 80) and lower and clearer smoke emission. Furthermore, phenolic based laminates are capable of complying with the adopted standard tests for the evaluation of the toxicity of smokes. For these reasons, where low flame spread and the smoke emission are primary constraints (railway and underground carriages), hybrid laminates are often employed, in which the internal structural layers are polyester based while the external (fire-resistant) ones are phenolic based.

The use of fire retardant additives and coating has been found to cause an increase of both the smoke emission and the toxicity of gases. For these reasons these agents have been rejected from all the major naval construction (i.e. minehunters).

PVC foam which is widely used in marine sandwich structures decomposes at temperatures over 200°C with emission of HCl gas which is toxic and highly corrosive. For this reason it has been embargoed by several navies.

3.5 REFERENCES

1] Pritchard, G., Speake, S.D., "The Use of Water Absorption Kinetics Data to Predict Laminate Property Changes", Composites, 18 (3), July 1987. pp 227-232.

2] Springer, G.S., "Environmental effects on Composite Materials", Technomics, Lancaster, PA, 1984.

3] Hoffer, K.E., Skaper, G.N., "Effect of Marine Environment on Bending Fatigue Resistance of Glass Cloth Laminates", Proc. Intl. Conf. *Marine Applications of Composite Materials*, Florida Institute of Technology, Melbourne, 1986.

4] Jacquemet, R., Lagrange, A., Grospierre, A., Lemascon, A., "Etude du Comportement au Vieillissement de Stratifies Polyester/Verre E en Milieu Marin", (in French), Proc. 1st Conf. *La Construction Navale en Composite*, IFREMER, Nantes, 1988.

5] Morii, T., Tanimoto, T., Maekawa, Z., Hamada, H., Yokoyama, A., Kiyosumi K., Hirano, T., "Effect of Water Temperature on Hygrothermal Aging of GFRP Panel", Proc. 6th Intl. Conf. *Composites Structures*, Elsevier Applied Science, London, 1991.

6] Gibbs & Cox, Inc., "Marine Design Manual for Fibreglass Reinforced Plastics", McGraw Hill, New York, 1960.

7] Mortaigne, B., Hoarau, P.A., Bellenger, V., Verdu, J., "Water Sorption and Diffusion in Unsaturated Polyester Networks", (in

French), Revue Scientifique et Technique de la defence, DGA, Paris, 1991.

8] Dixon, R.H., Ramsey B.W., Usher, P.J., "Design and Build of the GRP Hull of HMS Wilton", Proc. Symp. *GRP Ship Construction*, RINA, London, October 1972.

9] Owens-Corning Fibreglass, Fiber Glass Marine Laminates, 20 Years of Proven Durability, Toledo, OH, 1972.

10] Fried, N., Graner, W.R., "Durability of Reinforced-Plastic Structural Materials in Marine Service", Mar. Tech., July 1966. pp 321-327.

11] Scott, R.J., "Fibreglass Boat Design and Construction", Science Press Inc., New York, 1973.

12] Norwood, L.S., "Blister Formation in Glass Fibre Reinforced Plastic: Prevention rather than Cure", Proc. 1st Intl. Conf. *Polymers in a Marine Environment*, IMarE, London, 1984.

13] Cook, Polycor Polyester Gel Coats and Resins.

14] Corrado, G., "Il Fenomeno dell'Osmosi nella Nautica: Probabili Cause e Metodi di Difesa Preventivi", (in Italian), AITIVA Conf. 30th Genoa Boat Show, Genova, 1990.

15] Norwood, L.S., Holton, E.C., "Marine Grade Polyester Resin for Boat Building in the 1990s", The Society of the Plastics Industry, London, 1991.

16] Kokarakis, J., Taylor, R., "Theoretical and Experimental Investigation of Blistered Fibreglass Boats", Proc. 3rd Intl. Conf. *Marine Applications of Composite Materials*, Florida Institute of Technology, Melbourne, 1990.

17] Smith, C.S., "Design of Marine Structures in Composite Materials", Elsevier Applied Science, London, 1990.

18] Kasen, M.B., "Mechanical and Thermal Properties of Filamentary Reinforced Structural Composites at Cryogenic Temperatures", Cryogenics, December 1975.

19] Schram, R.E., Kasen, M.B., "Cryogenic Material Properties of Boron, Graphite and Glass Reinforced Composites", Mater. Sci. Engg., **30**, 1977. pp 197-204.

20] Amoco Chemical Company, "25 Year Old Tank Showcases Corrosion Resistance of Isopolyesters", Amoco Bulletin IP-88a, 1988.

21] Chemical Resistance, Crystic Mongraph No. 1, Scott Bader Co. Ltd, 1971.

22] Derakane Vinyl Ester Resins: Chemical Resistance Guide, Dow Chemical Europe, Horgen, Switzerland.

23] Ship Structures Committee, "Marine Composites - Investigation of Fibreglass Reinforced Plastic in Marine Structures", SSC Report 360, U.S. Coast Guard, Washington, DC, 1990.

24] U.S. Code of Federal Regulations 29 CFR 19100.1000.

25] Occupational Safety and Health Administration, Air Contaminant Standard, OSHA, 1989.

26] International Maritime Organization, Int. Convention on Safety of Life At Sea, IMO, London, 1974/83.

27] Morchat, R.M., Allison, D.M., Marchand, A.J., "Large Scale Fire Performance Testing of Composite Structures", J. Mar. Struct., 4 (2), 1991.

28] Ulfvarson, A.Y.J., "Superstructures of Large Ships and Floating Offshore Platforms Built in FRP Sandwich - A Feasibility Study", Proc. 1st Intl. Conf. *Sandwich Constructions*, Royal Institute of Technology, Stockholm, 1989.

29] Smith, C.S., Chalmers, D.W., "Design of Ship Superstructures in Fibre Reinforced Plastic", Trans. RINA, **129**, 1987. p 45.

30] International Maritime Organization, Materials Other Than Steel for Pipes - Information Note FP 35/WP.9, IMO, London, 1990.

31] International Maritime Organization, Code of Safety for Dynamically Supported Craft - IMO resolution, A.373 (X), IMO, London, 1977.

32] International Convention for the Safety of Fishing Vessels, Torremolinos 1977.

33] Puccini, G., Porcari, R., Cau, C., "Experimental Activity on GRP Ship Superstructures Panels", Proc. Intl. Symp. *Ship and Shipping Research*, Genova, 1992.

34] Mableson, A.R., Osborn, R.J., Nixon, J.A., "Structural Use of Polymeric Composites in Ship and Offshore", Proc. 2nd Intl. Conf. *Polymers in a Marine Environment*, IMarE, London, 1987.

4 PRODUCTION BY RESIN TRANSFER MOULDING

4.1 INTRODUCTION

At the introduction of the reinforced plastic products just after World War II the hand lay-up technique was used for the production of those products. This technique has persisted as the years passed on because it is a simple, handcrafted technique with low cost tooling. As with all handcrafted techniques, the quality of the products depends on the individual craftsmanship of the operator. Even with good craftsmanship the products only have one side with defined geometry, the thickness of the products is not well controlled and consequently the fibre-volume fraction can only be guaranteed within a wide range. Moreover the material will contain a considerable number of air bubbles, or voids.

The technique is time consuming and consequently wage expenditure is the greater part of the cost price of the product and the cycle time for moulding is rather long. For volume manufacture of a product several moulds have to be used to achieve sufficient productivity. As the years passed attempts were made both to improve the productivity of the process and to raise the quality of the product.

A first attempt to raise the productivity and to lower labour costs was the spray lay-up technique but this was achieved at the cost of lower mechanical properties. The loss of mechanical properties is due to the fact that with this technique a random fibre distribution in all directions is obtained and tailoring the anisotropy of the material to the loading of the product is not possible. To improve the quality of the product and to achieve a higher fibre-volume fraction the following three techniques were developed [1]:

a. Bag moulding
b. Drape moulding
c. Autoclave moulding

In all these techniques, after hand lay-up or spray lay-up, the surplus of resin is pressed out of the material. The resin flowing out of the product will drag the air bubbles along, but as the air bubbles flow with a higher velocity than the resin, a part of the bubbles in the product will be removed. When the geometry of the product has to be controlled completely, a mould consisting minimally of two parts has to be used. The cavity of the mould has the geometry of the product. The processing techniques based on this principle form a group called matched die moulding. One of these techniques is Resin Transfer Moulding (RTM) and it will be discussed in more detail in the rest of this Chapter.

4.2 THE RESIN TRANSFER MOULDING TECHNIQUE

4.2.1 Description of the RTM Technique
Dry fibre material is placed in the mould in an open position. After closing the mould the resin is injected in the mould cavity, see figure 4.1.

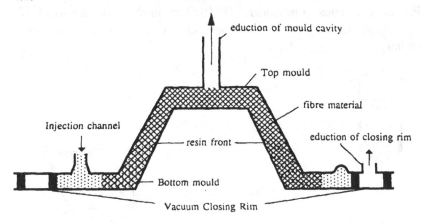

Figure 4.1. Resin transfer moulding technique in outline.

The resin flows from the injection channel through the fibre material to the outlet channel. As the fibre material resists the flow a driving force is needed to obtain the contemplated flow. The driving force is a result of the difference between liquid pressure in the injection channel and the air pressure in the mould.

Essentially the filling process replaces the air in the fibre material by the fluid. It occurs more easily in the voids between the fibre bundles than in the microvoids between the fibres in the bundle.

Consequently air is trapped in the microvoids when the flow front progresses. The air in the microvoids escapes later and forms air bubbles in the resin between the bundles and, having a higher velocity than the resin, is transported towards the flow front.

With the RTM process, products requiring tight tolerances and both inside and outside surfaces smooth, can be produced. The fibre-volume fraction is fairly good and the void content is low. Moreover, the concentration of fumes in the air of the workshop is reduced considerably compared with the hand lay-up technique. To achieve a short-time for filling the mould for a product with a high fibre-volume fraction, a large pressure difference between injection pressure and air pressure in the mould is necessary. The high injection pressure requires strong and rigid moulds resulting in high investment costs and consequently this method can only be used for mass production of products such as parts for the automotive industry.

4.2.2 Vacuum Injection Technique
For smaller series of products (100-1000 in number, say) a variant of the RTM technique is developed called the vacuum injection technique, see figure 4.2.

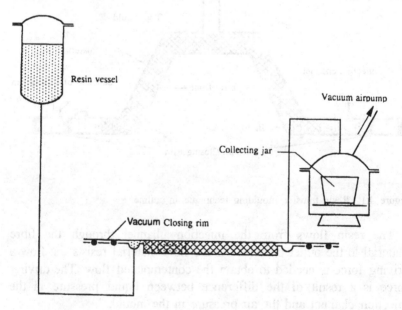

Figure 4.2. Vacuum Injection Technique in outline.

In this technique the driving force to transport the resin through the fibre material is obtained by a low air pressure in the mould (mostly 600-700 mbar absolute). The injection pressure is only a little higher than the atmospheric pressure (1-2 bar) so that light fibre reinforced plastic moulds can be used. These moulds are relatively inexpensive and can be produced by the fibre reinforced plastic industries themselves. Because the fumes are discharged by the vacuum pump they can be treated in a separate process reducing the environmental load of the air in the surroundings of the plant considerably.

To prevent air leakage into the mould cavity, an encircling annular channel, called a closing rim, is present between the mould parts. In the closing rim, the air pressure is always lower than the pressure in the mould cavity in order to prevent air leakage into the cavity. The vacuum in the closing rim presses the two parts of the mould on each other.

As a consequence of the stricter legislation for the environment in the workshop (in countries such as The Netherlands), more and more hand lay-up industries are switching to the vacuum injection technique, since this is the least expensive variant of the RTM techniques.

As the filling time of the mould is determined by the flow velocity of the resin passing through the fibre material, the relation between the factors influencing the flow have to be considered.

4.3 FLUID DYNAMICS OF THE FLOW OF A LIQUID THROUGH A POROUS MATERIAL

4.3.1 Theoretical Foundations [2,3]

The velocity of the flow of the liquid through a porous material depends on:

- the pressure gradient in the flow direction
- the viscosity of the liquid
- the porosity of the porous material

and is given by the Darcy Law:

$$q_x = \frac{K}{\mu} \frac{dp}{dx} \tag{4.1}$$

found in research on geological problems. Here, q_x is specific volume output in the x-direction, μ is dynamic viscosity of the fluid and K is permeability of the porous medium.

The permeability of a porous medium depends on its structure. For porous media consisting of particles as grains, it is characterised by the concepts of porosity and specific wetted surface.

The porosity ψ is defined as the ratio of the pore volume to the total volume. The specific wetted surface S is equal to the total surface of the outside of the grains per unit of volume of those grains. For geological underground water flow, Kozeny has found that the permeability is given by:

$$K = \frac{1}{C} \frac{\psi^3}{S^2 (1 - \psi)^2}. \tag{4.2}$$

The dimensionless constant C has to be determined experimentally. Carman has found in his experiments with different grainlike media, values of about 5 for the constant C. The equation is called the Kozeny-Carman equation.

The specific volume output has the dimension of a velocity, but as it is related to the total surface of the cross-section of the flow, is not equal to the real fluid velocity. For the prediction of the flow-front-velocity when filling a mould with the RTM technique, the real average velocity has to be used.

4.3.2 Flow-Front-Velocity
As the pore surface is a fraction ψ of the total surface perpendicular to the flow direction, the liquid real average velocity is:

$$V = \frac{q_x}{\psi}. \tag{4.3}$$

To predict the flow, front velocity has to be used.

$$V = \frac{K}{\mu} \frac{dp}{dx} \quad \text{with} \quad k = \frac{K}{\psi} = \frac{1}{C} \frac{\psi^2}{S^2 (1 - \psi)^2}. \tag{4.4}$$

k will be called flow factor, as the process is considered from the viewpoint of the liquid. Having discussed the fluid dynamics of the flow of the resin filling the mould, the filling process will be discussed in more detail.

4.4 MOULD FILLING PROCESS
As already explained, the essence of the process is to replace the air between the fibres by resin. Because the reinforcing fibres are bundled together in rovings or yarn, there are two kinds of voids in the material, see figure 4.3.

a. Voids between the fibre bundles, having dimensions more or less equal to the dimension of the cross-section of the bundle, in the order of magnitude of a few millimetres.

b. Microvoids between the fibres in one bundle having dimensions corresponding with those of the fibre cross-section, in the order of magnitude of a few micrometers.

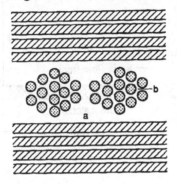

Figure 4.3. Cross section of right angle crossing of fibre bundles with voids between the bundles and microvoids between fibres in the bundles.

In this Chapter, filling the first type of voids is called "filling of the mould" and filling the second type of voids is called "impregnating the fibre bundles".

4.4.1 Flow of Resin Between Fibre Bundles
First, filling the mould for a strip of fibre reinforced material is discussed. The mould is considered rigid and is as shown in figure 4.4.

When the flow front is at a position between inlet and outlet gates, in every cross-section of the filled part of the mould the same volume of resin is passing at the same time. As the permeability of the fibre material and the viscosity of the resin is assumed to be constant the pressure gradient must be constant, see figure 4.5.

As the flow front progresses, the distance between injection gate and flow front increases; the pressure gradient is, as injection and air pressure in the mould are constant, inversely proportional to that distance. Consequently, the flow-front-velocity is also inversely proportional to the distance between flow front and injection gate. After the flow front has passed a fixed point in the mould the liquid pressure at that point increases, at first rather rapidly but later on more slowly.

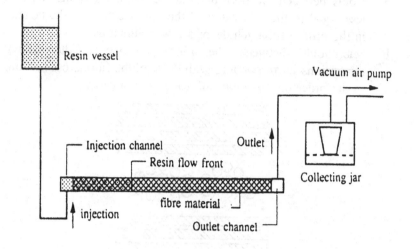

Figure 4.4. Simplified outline of the vacuum injection moulding of a flat strip fibre reinforced material.

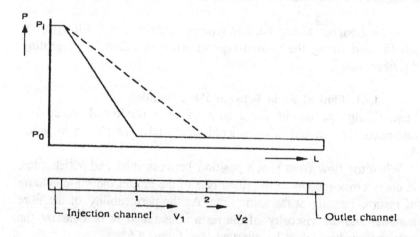

Figure 4.5. Pressure curve and flow fron velocity in dependence of the position of the flow front in a straight channel flow.

In figure 4.6 the pressure time curve for a point close to the injection gate is given. All these phenomena are influenced by the impregnation of the fibre bundles which is discussed in section 4.4.2.

A second simple example of a product would be a circular plate injected at the centre of the mould with an outlet gate along the

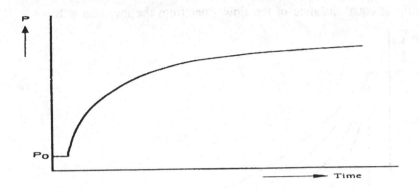

Figure 4.6. The fluid pressure at a distinct point near the injection gate as function of time; rapid pressure rise after the fluid flow front reaches that point.

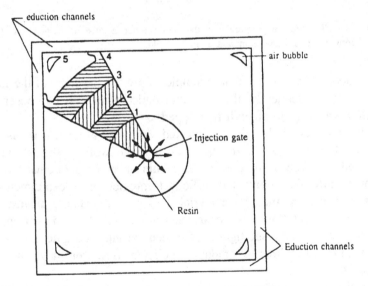

Figure 4.7. Central injection of a quadrangular plate, eduction along the sides of the plate.

cylindrical wall of the mould. The flow front will be circular seen from above the mould. The larger the radius of the circle the larger is the flow front surface, see figure 4.7. Because in each circular surface of the filled part of the mould the same volume of liquid will pass, the velocity of the liquid will decrease in inverse proportion to the distance from the injection gate.

The pressure time curve is given in figure 4.8. The pressure gradient at the flow front is small compared with the gradient for the strip at equal distance of the flow front from the injection gate.

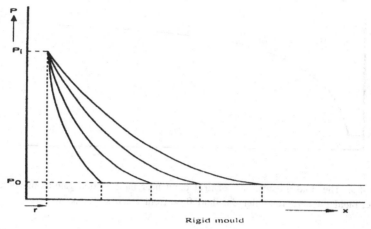

Figure 4.8. The drop of pressure over the mould in dependence of the position of the flow front, at central injection and a rigid mould.

In both cases discussed it is simple to predict the form of the flow front and consequently the inlet and outlet gates can be chosen in such a way that no air will be trapped.

When a rectangular plate is injected via a central gate, it is easy to predict the front up to the moment that the wall of the mould is reached. Viewed from above the mould it will be circular. What happens after that moment is difficult to predict. From experiments it is known that, as the fibre material does not fit perfectly against the wall, a short-cut flow will occur and voids can be formed even in the corners of the plate, see figure 4.7. When the injection gate is situated at the sides of the mould, gradually a circular flow front develops, see figure 4.9.

In figure 4.10 the pressure time curve is given; it is concave as the velocity of the resin near the centre outlet gate is high compared with the velocity near the inlet gate. The injection time using circumferential injection is shorter than with central injection. It is obvious that prediction of the flow front for more complex shaped products is difficult and consequently the positions of inlet and outlet gates, which guarantee void-free production, is also difficult to predict.

In the author's laboratory a computer simulation of the filling process has been developed to facilitate the design of the mould and is discussed in section 4.7.

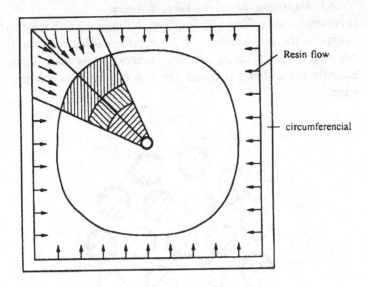

Figure 4.9. Central eduction of a rectangular plate with injection round all sides.

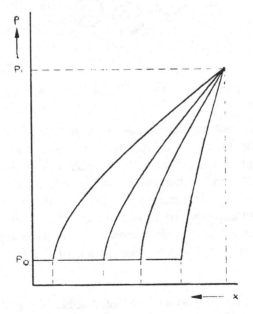

Figure 4.10. The drop of pressure over the mould.

Apart from filling the mould in such a way that no voids will occur, impregnation of the bundles is essential for production of high-quality products.

4.4.2 Impregnation of the Fibre Bundles

The cross-section of a fibre bundle given in figure 4.11 shows that the dimensions of the cross-section of the microvoids between the fibres are very small. The diameter of a fibre is mostly less than 10 μm and consequently the distance between fibres in the cross-section are only a few μm.

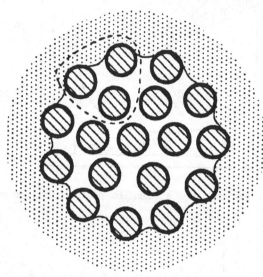

Figure 4.11. Simplified outline of the cross section of a fibre bundle enveloped by resin before impregnation.

It is obvious that capillary action is an important driving force to impregnate the bundle, but also that the flow resistance in the capillary is high. RTM experiments in a glass mould showed that the resin flows between the bundles before the penetration of the resin into the microvoids of the bundle. The fibre bundles are enveloped by resin before the impregnation of the bundles starts. Consequently, the resin penetrates a fibre bundle perpendicular to the bundle axis. The phenomena acting during the radial afflux of the fibre bundle are discussed below.

4.4.2.1 Capillary action with radial afflux of the fibre bundles [4-6]

Capillary action is caused by the difference in surface energy of the solid and the liquid. When a drop of liquid is laid on a horizontal plate of a solid material the plate is more or less wettened. In figure 4.12 the cross-section of the drop is given and shows a distinct contact angle between the surface of the liquid and the plate. The smaller the contact angle the better the wetting. As is generally known in a capillary tube the liquid having a small contact angle will form a concave curved liquid surface. In a vertically placed capillary tube the liquid rises in the tube up to the moment that the surface tension forces are equal to the weight of the liquid in the tube above the liquid surface outside the tube.

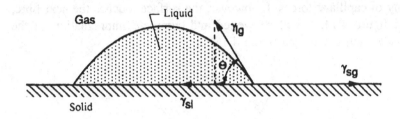

Figure 4.12. A drop of a liquid on a horizontal solid plane. Equilibrium of forces in horizontal direction at the edge (contact angle θ).

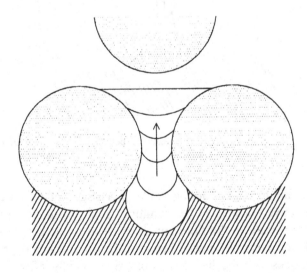

Figure 4.13. Changing meniscus at a distance between fibres of ¼ of the fibre diameter.

When penetrating a fibre bundle radially the dotted area of the cross-section of the fibre bundle is considered, see figure 4.11 and figure 4.13. When the liquid comes in contact with the two fibres in the outside surface of the bundle, a contact angle between the liquid surface of the micro front and the surface of the fibres develops and a meniscus is formed. Through capillary action the meniscus is pulled into the slot between the fibres. The flow of the liquid in the very narrow slot experiences a high resistance but slowly an equilibrium position is reached, being a flat surface.

When, as in figure 4.13, this surface is not in contact with the next fibre in the fibrestack, no further penetration can occur as a result only of capillary forces. If however, the surface touches the next fibre, see figure 4.14, it will be wetted and complete impregnation of the bundle will occur, see figure 4.15.

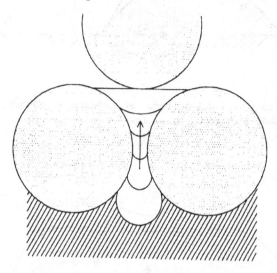

Figure 4.14.   Changing meniscus at a distance between fibres of $^1/_8$ of the fibre diameter.

In radial impregnation during RTM processes not only capillary forces are active but also forces resulting from pressures. Assuming that the air pressure in the fibre bundle is equal to the air pressure in the mould, there is a difference between the air and liquid pressures at the meniscus. As the liquid pressure rises rapidly after the front passage the meniscus is "blown up" and becomes convex, see figure 4.16. When it touches the next fibres in the cross-section further impregnation starts. There is a time lapse between the passing of the macro front and the complete impregnation of the bundle. Impregnation has so far been discussed for one cross-section of one fibre bundle in RTM processes, however there are many bundles with varying cross-sections.

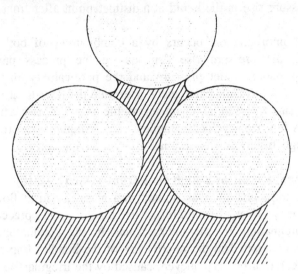

Figure 4.15. Meniscus after touching the fibre of the second circular ring of fibres in the cross section of the bundle.

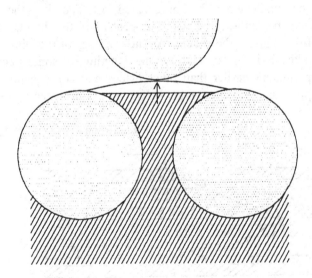

Figure 4.16. Blowing up the meniscus by pressure rise after flow front passage.

4.4.2.2 Impregnation during RTM processes
As discussed earlier, the driving forces for impregnation are

- Capillary action and
- Pressure rise in the liquid at a distinct point after front passage

In general impregnation occurs by a combination of both driving forces. As the pressure rise depends on the process parameters injection pressure, outlet pressure and the permeability of the fibre material whilst the same parameters determine the flow-front-velocity, the latter can be used to characterise the pressure rise after front passage. A low front velocity implies a small liquid pressure driving force, a high front velocity implies a large driving force.

A. **Impregnation at a high flow-front-velocity [7-10].**
The impregnation was studied in a glass mould at flow-front-velocity over 10 mm/min. The macro front precedes the impregnation by the micro fronts in the bundle and air is always trapped in the microvoids of the bundle. Impregnation appeared to be very uneven, caused by the irregular stacking of the fibres in the bundle. When the fibres are stacked regularly, impregnation would be regular, as given in figure 4.17. In this case the micro fronts would form a hollow cone and the centre of the bundle would be impregnated last. The air in the bundle would be transported via the centre of the bundle to the unfilled part of the mould. As the stacking of the fibres in the bundles is irregular, and varying lengthwise, some places are impregnated earlier than other places and consequently air is trapped in some microvoids between the fibres, see figure 4.18.

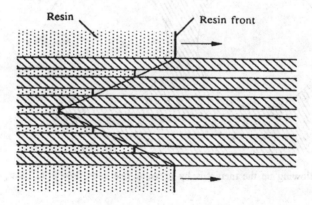

Figure 4.17. Impregnation of a regular fibre bundle in a radial direction.

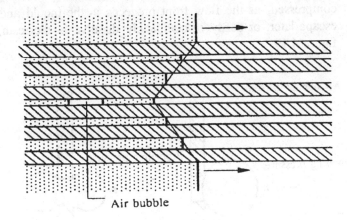

Air bubble

Figure 4.18. Impregnation of an irregular fibre bundle in radial direction with air voids.

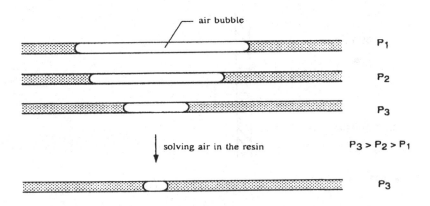

air bubble

P_1

P_2

P_3

solving air in the resin $P_3 > P_2 > P_1$

P_3

Figure 4.19. Compression of an air microvoid in a capillary manifests in shortening; likewise dissolving of the air in the resin at pressure.

As the liquid pressure rises rapidly after front passage the trapped air in the micro bubbles is compressed, see figure 4.19; it was observed that after compression quite a few micro bubbles escaped out of the bundle and were transported by the liquid. As there is a pressure gradient in the liquid there is a relative velocity of the air bubbles to the fluid. The air bubbles overtake the macro front or stop dead in narrow pores, see figure 4.20, of the fibre material. Those air bubbles are compressed, as the flow front proceeds in the mould and can escape later, or partly or completely dissolve in the resin, see figure 4.19.

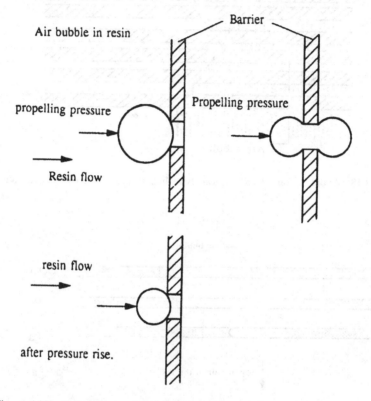

Figure 4.20. Deformation of an air bubble with compression.

The dissolving of the air out of the air bubbles was studied separately. The less the air in solution in the resin before injection, the more air can be absorbed. In the author's laboratory the resin in the resin reservoir is always deaerated before the injection starts to achieve a maximum airdissolving capacity before complete saturation with air is reached.

B. Impregnation at a low flow-front-velocity.
In this case impregnation occurs mainly by capillary action. The micro fronts in the bundle precedes the macro flow front, but due to the irregularity of the bundles more air is trapped than with high flow- front-velocity.
    Up till now it was assumed that the mould is rigid. The

vacuum injection technique moulds however are flexible and consequently deform during the filling process. As the cross-section of the flow changes with liquid pressure the pressure gradient changes at a distinct moment in the filled part of the mould to produce a strip.

4.5 THE INFLUENCE OF MOULD FLEXIBILITY ON THE FILLING PROCESS

The influence of the mould flexibility is discussed for the filling process of a mould for a rectangular plate with an injection gate along one side of the plate, and an outlet gate at the opposite side. The flow is linear so the process can be compared with the production of the strip in a rigid mould given in section 4.4.1.

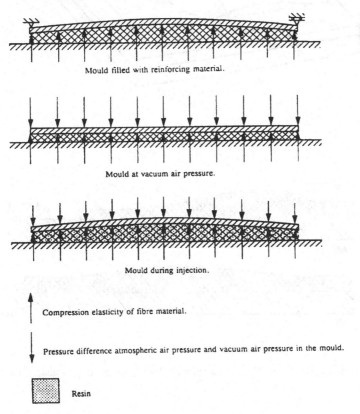

Mould filled with reinforcing material.

Mould at vacuum air pressure.

Mould during injection.

Compression elasticity of fibre material.

Pressure difference atmospheric air pressure and vacuum air pressure in the mould.

Resin

Figure 4.21. Three stages during vacuum injection with a flexible top mould.

The fibre material used in the vacuum injection technique is compressible and this has to be taken into account when discussing the filling process. In order to simplify the discussions the bottom mould is assumed to be rigid, as is the case in the author's laboratory, where a glass plate is used. The top mould is of fibre reinforced plastic material and flexible. When after placing the fibre reinforcing material the mould is closed, the compression elasticity of the fibre material exerts an upwards force on the top mould, causing a more or less spherical bulging of it, see figure 4.21.

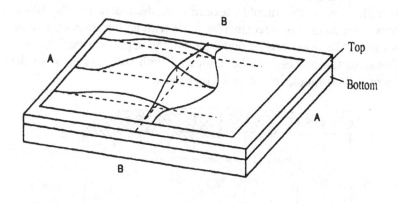

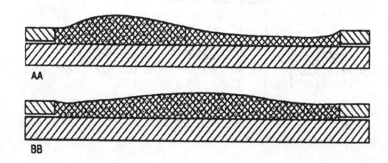

Figure 4.22. The filling of a flat mould with flexible top mould, bulging of the filled part.

When the vacuum connection of the mould is opened a downward force is developed. The vacuum pressure can be adjusted in such a way that the top mould is flattened perfectly. However when the injection starts the liquid pressure on the top mould bulges the filled

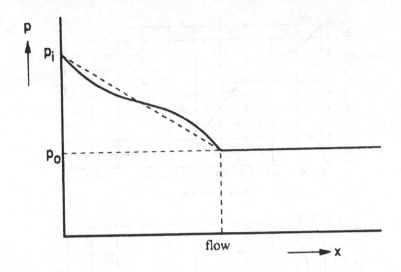

Figure 4.23. The pressure curve when the top mould part bulges.

part of the mould, see figures 4.21 and 4.22. The pressure curve for
the central line of the mould is given in figure 4.23 as a full line,
whereas the dotted curve gives the pressure curve for a rigid mould.
As the same amount of resin passes each cross-section of the bulged
part of the mould, the velocity in the bulged part is lower than the
front velocity. The pressure gradient in the bulged part is smaller than
at the front, and the front velocity in the flexible mould is larger than
in the rigid mould. For studying the flow in a flexible mould, a
computer simulation has been developed, using an experimentally
determined compression curve of the fibre material and the calculated
flexibility of the mould. In figure 4.24 both the pressure curve and the
bending of the mould are shown.

Apart from a higher flow front speed in a flexible mould the
vacuum injection technique gives a better surface of the product than
with rigid moulds. The flexibility of a mould and the low pressure in
the mould results in an evenly spread compacting pressure of the
mould on the product compensating the reaction shrinkage of the
resin, which, with a polyester resin is up to 9% of volume.

As the filling process depends to a high degree on the permeability
of the fibre material the permeability of those materials has to be
discussed in more detail.

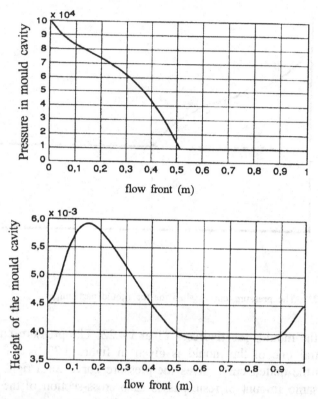

Figure 4.24. Pressure curve and bending of the flexible top part of the mould at a distinct moment during the moulding of a rectangular plate with sides of 1 m.

4.6 PERMEABILITY OF FIBRE MATERIALS

In section 4.3.1 (equation 4.2), the Kozeny-Carman relation is given; to calculate the flow factor it must be kept in mind that those materials contain bundles of fibres and that in filling the mould the resin flows between those bundles. To calculate the permeability and specific wetted surface of a fibre material the bundles can be conceived as solid voids having a circular cross-section which envelopes the outer layer of fibres, see figure 4.11. The Carman constant can only be found experimentally, with a special measuring mould such as one in the author's laboratory.

4.6.1 Permeability Measuring Mould

The measuring mould consists of a glass plate and an aluminium plate separated by a spacer forming a rectangular flow channel, see figure 4.25. During the permeability measurement the quantity of liquid flowing into the mould, the liquid pressure at five pressure sensors on the centre line of the bottom mould with 200 mm pitch and the pressure in resin vessel and overflow vessel are registered. The

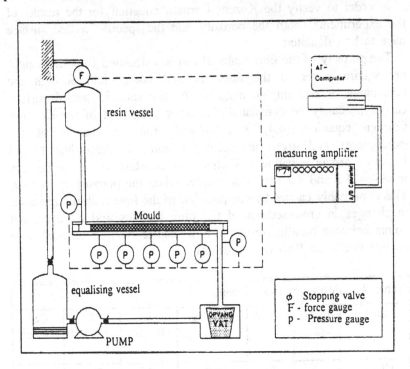

Figure 4.25. Outline of the measuring mould.

electronic signals from all the measuring sensors are stored in a computer followed by calculation of the permeability and on line presentation on a screen. To obtain accurate and reliable results the placing of the fibre material must be such that no short-cut flow along the walls of the mould can occur. This is achieved by cutting the fibre material 1 mm broader than the width of the measuring channel. For the measurement silicon oil with two different viscosities (20-200 mPaS) is used having a constant viscosity over a long period of time.

The permeability is calculated both during filling of the measuring mould and the continuous flow after the mould is completely filled.

4.6.2 Permeability of Different Types of Fibre Material
First of all the mould was filled with one type of fibre material, mats or woven fabrics, measuring at different flow levels obtained by

changing the difference between injection pressure and air pressure in the overflow vessel, and at different fibre fractions by changing the number of layers of fibre material in the mould.

In order to verify the Kozeny-Carman equation for the results of the experiments, both the porosity and the specific wetted surface have to be calculated.

The porosity of the fibre material can be calculated from the weight per square meter of the fibre material taking into account the percentage of size on the material. As the specific wetted surface cannot so easily be calculated the factor $1/CS^2$ out of the Kozeny-Carmen equation (4.4) is calculated from the results of the experiments and given in tables 4.1 and 4.2. According to the Kozeny-Carman equation CS^2 should be constant but it can be seen in tables 4.1 and 4.2 that it is smaller when the porosity is smaller. This is probably caused by compression of the fibre material resulting in changes in cross-section of the fibre bundles and more contact points between bundles resulting in a smaller "wetted surface" as at contact points no flow of liquids occurs.

Table 4.1. Permeabilities of mats.						
Material	Aantal lagen	v_f [%]	porosity ψ_β	k $[10^{-9}\,m^2]$	K $[10^{-9}\,m^2]$	CS^2 $[10^9\,m^2]$
Unifilo U816 (450 g/m²)	2	7.3	0.909	24.8	26.8	3.63
	3	11.8	0.852	9.43	10.7	3.02
	4	15.8	0.802	4.32	5.13	3.07
Unifilo U720 (450 g/m²)	3	11.3	0.859	13.1	14.8	2.42
	4	15.1	0.811	6.50	7.66	2.31
Unifilo U850 (450 g/m²)	3	11.8	0.852	15.2	17.2	1.87
	4	14.4	0.820	8.72	10.2	1.95
Conformat N751 (450 g/m²)	2	6.9	0.914	4.07	4.37	25.2
	3	10.1	0.874	2.27	2.52	18.4
Conformat N754 (450 g/m²)	2	6.3	0.921	13.2	14.1	9.55
	3	9.4	0.882	6.25	6.90	7.97

Material	Aantal lagen	v_f [%]	porosity ψ_β	k $[10^{-9} \, m^2]$	K $[10^{-9} \, m^2]$	CS^2 $[10^9 \, m^2]$
Cotech (1200 g/m²)	4	38.9	0.514	1.28	2.09	0.45
	5	48.2	0.398	0.11	0.212	1.57
Dubbelkeper (830 g/m²)	5	31.9	0.601	0.416	0.611	3.29
	6	37.8	0.528	0.122	0.176	5.39
Frencken W2439 (395 g/m²)	15	43.7	0.454	0.278	0.494	1.13
	17	50.1	0.374	0.076	0.152	1.75
	19	58.3	0.271	0.024	0.058	1.57
Injectex (295 g/m²) ‡	9	20.2	0.748	5.12	6.42	1.28
	11	24.8	0.690	3.46	4.60	0.99
	14	33.0	0.588	0.59	0.88	2.02
Injectex (295 g/m²) §	9	21.0	0.738	2.01	2.54	2.90
	11	24.4	0.695	0.92	1.22	3.92

Table 4.2. Permeabilities of woven fabrics.

‡. Flow along the twisted warp yarns
§. Flow along the weft yarns

From the results of the experiments it is found that there are big differences in permeability for different fibre materials. As in most products a mix of mats and woven fabrics are used, those combinations have also to be tested.

4.6.3 Permeability of Combinations of Mats and Woven Fabrics

As the permeabilities of mats and woven fabrics differ considerably, the flow of the liquid in the mats has a higher velocity than in the woven fabrics. At a fixed moment the flow front in the mats is preceding the flow front in the woven fabrics, see figure 4.26; consequently the flow in woven fabric is perpendicular to the sheet of woven fabric.

This process is modelled both analytically and numerically, with both approaches giving the same results. Sometime after the start of the injection, the flow front in the mats precedes the filling of the woven fabric with a constant distance. The flow-front-velocity is determined by the permeability of the mats. In modelling a combination of mats and woven fabrics the compressibility of both materials has to be taken into account, being far greater for the mats than for the woven fabrics.

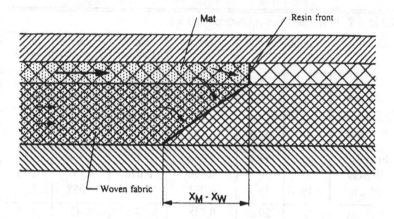

Figure 4.26. Model of the flow for a mat - woven fabric combination.

When closing the mould the compression mainly appears in the mats and consequently the permeability of the mats is diminished far more than of the woven fabrics. The compression curves of mats and woven fabrics have to be determined experimentally.

As the compression pressure in the closed mould is the same for the mats and the woven fabrics and the total height of the mould cavity is known, the thickness of both materials in the mould can be calculated.

From the experiments it can be concluded that in a combination of mats and woven fabrics the mat is acting as the "flow layer".

As in professionally designed products the mechanical properties of the material are tailored to the load on it, the permeability differs locally. In the design of the mould this has to be taken into account. Mostly this is done using the experience of the designer, but it is difficult to guarantee that inlet and outlet gates are placed in such a way that no voids will occur. When in the first experiments with a new mould voids occur, the mould design has to be changed. This is time consuming and costly, so computer simulation for complicated products is very useful.

4.7 FLOW FRONT SIMULATION

4.7.1 Description of a Simulation Technique
In the filled part of the mould the Darcy equation is assumed to be valid as well as the continuity equation. Both equations together describe the distribution of the liquid velocity and the liquid pressure

over the filled part of the mould.

At the inlet gate, the injection pressure and at the flow front the air pressure in the mould cavity, are prescribed. At the walls of the mould the fluid velocity perpendicular to the wall is presumed zero. Solution of both equations is done via a variant calculation using a finite element grid generated by a mesh-generator, resulting in a computer simulation programme.

When using this programme, data has to be entered for:

- the geometry of the mould cavity
- the geometrical position of the inlet gate
- the distribution of the permeability of the fibre material over the mould

The flow front is given by the programme at different elapsed times from the start of the injection.

To show the effectiveness of the programme a few simulations will be discussed.

4.7.2 Simulations of the Filling of a Flat Plate of Fibre Material with a Large Ratio of Permeabilities in Different Directions (Anisotropy)

The permeability of the fibre material in two perpendicular directions has a ratio of 10-1. One of the directions is at 60° to the horizontal side of the plate. The result of the simulation is given in figure 4.27 showing an elliptical flow front.

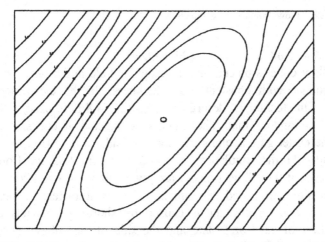

Figure 4.27. Injection of a rectangular plate with anisotropic fibre material.

4.7.3 Simulation of the Filling of a Flat Plate with Large Differences in Permeability of the Fibre Material

The rectangular flat plate is divided in four equal parts, two parts filled with random mats, the other two parts with woven fabric, see figure 4.28. The ratio of the permeabilities of the fibre materials was 9-1. The computer simulation is given in figure 4.29. The result of the experiment in figure 4.28, shows good correlation between simulation and experiment.

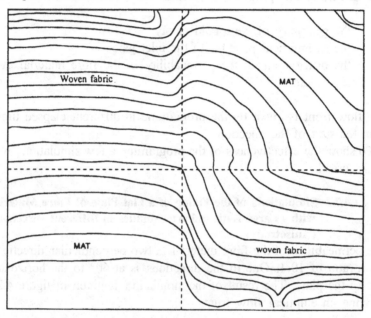

Figure 4.28. Flow fronts during injection.

4.7.4 Hollow Cube on a Flat Plate

The hollow cube is placed on a rectangular hole in the plate resulting in a product with constant wall thickness. In the experiment the vacuum film technique described in section 4.8.1, is used, the fibre material is random mat. The product is injected at one side of the plate, the outlet gate is along the opposite side of the plate. In figures 4.30 and 4.31 the simulation is given, in figure 4.32 the front and back views of the filling process are given, showing good correlation between simulation and experiment. The slight difference between simulation and experiment is caused by short-cut channels at the intersection of cube and plate and the side of the cubes, caused by compression of the fibre material in bending.

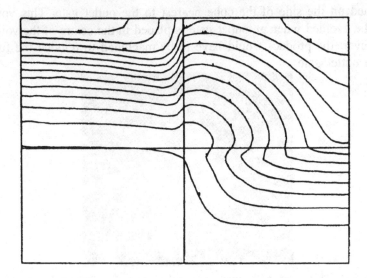

Figure 4.29. Computer simulation.

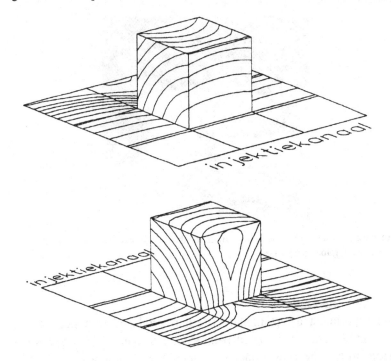

Figure 4.30. Flow front of a hollow cube (front and back view).

As can be seen in both figures 4.31 and 4.32, a rather large void is formed on the side of the cube nearest to the outlet gate. This void can be avoided when an outlet gate is placed in the centre of the void, however, the product would have to be machined to remove the full resin outlet gate.

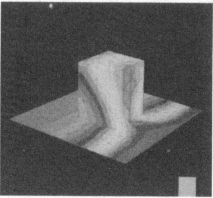

Figure 4.31. Photograph of the simulation on the screen of the monitor of a personal computer of the filling of a hollow cube.

Figure 4.32. Photograph of a distinct moment during the experimental filling of the hollow cube for the ideal case.

4.7.5 Rectangular Cistern

In the experiment a rectangular cistern is produced in a steel mould and injected at the vertical left hand side, see figure 4.33. As it is not possible to see the flow front in the mould, injection was stopped at different stages of the filling process. After curing the resin, partly filled products are obtained giving by approximation the geometry of the flow front. In this experiment the simulation given in figure 4.33,

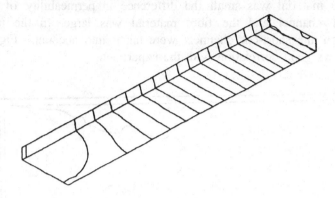

Figure 4.33. Flow front simulation for the ideal case.

and the experiment in figures 4.34, 4.35, 4.36 do not match. The difference was caused by short-cut channels at the edges of the fibre

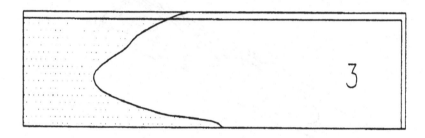

Figure 4.34. Resin front for the partly injected and cured mould.

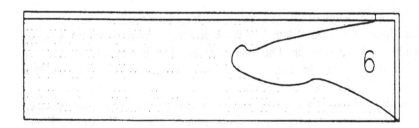

Figure 4.35. Resin front for the partly injected and cured mould.

material as it did not fit ideally in the mould. As the permeability of the fibre material was small the difference in permeability of the short-cut channel and the fibre material was large. In the next simulation the short-cut channels were taken into account. Figure 4.37 shows good correlation with the experiment.

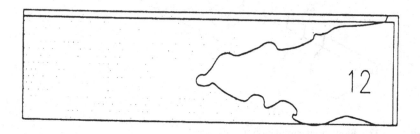

Figure 4.36. Resin front for the partly injected and cured mould.

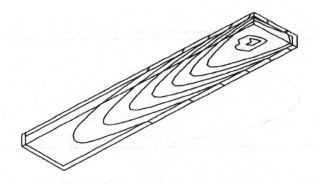

Figure 4.37. Computer simulation taking into account the short cut channels.

4.7.6 Door of an Automatic Coffee Machine

The door of the automatic coffee machine, shown in figure 4.38 is produced by a firm in The Netherlands. The height of the door is 1.6 m, out of one mould, three doors are produced daily and the injection gate is circumferential.

As one of the sides of the mould is transparent the filling process can be observed visually. The simulation is given in figure 4.39 and is in good conformity with the experiments. It can be concluded that:

-    A good conformity between computer simulation and reality exists

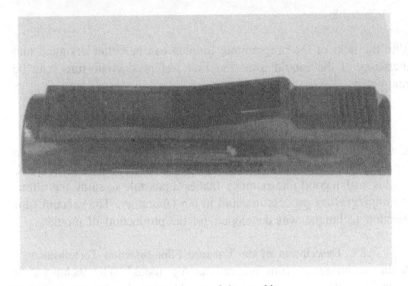

Figure 4.38. The product discharged out of the mould.

Figure 4.39. Simulation of the flow front geometry.

- Variable permeabilities of fibre material can be included in the programme
- Short-cut channels can be modelled in the programme, providing a good estimation of the channels can be made from experience with similar products
- Simulation of products having a complicated geometry is

possible

With the help of the programme, moulds can be better designed and alteration of the mould after the first test production runs, can be minimised.

4.8 PRODUCTION OF MOULDS WITH THE VACUUM FILM INJECTION TECHNIQUE

Moulds for the RTM techniques are produced traditionally with the hand lay-up technique. As for the research of RTM techniques, moulds with a good transparency makes it possible to study the filling and impregnation process in detail in our laboratory. The vacuum film injection technique was developed for the production of moulds.

4.8.1 Description of the Vacuum Film Injection Technique

In this technique a film is used as top mould. The technique is discussed in describing the production of the moulds for a scale model of a surfboard which was the research project of one of the author's student's.

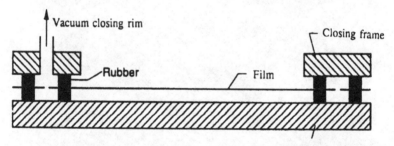

Figure 4.40. Outline of the film vacuum injection technique.

As only a few months are normally available for a student research project, the geometry of the scale model surfboard was simplified to facilitate the production of a positive model of it. As a result of this simplification one of the mould parts was a simple flat plate. As only one side of the plate mould has to be flat and smooth a simple steel plate flat table was used as "Bottom mould" for the vacuum film injection technique, see figure 4.40. Along the perimeter of the table two rubber strips are glued onto the table. On these rubber strips a film was placed, and on top of the film a closing frame with geometrical identical rubber strips. The two sets of rubber strips form a vacuum closing rim and the space inclosed by the table, the film and the rubber strip on the table formed the production cavity.

4.8.2 Production of a Flat Plate with the Help of the Vacuum Film Injection Technique

In the table two holes were drilled and injection and outlet pipe connection pieces were fitted in the holes, for the connection of injection vessel and overflow vessel to the mould by flexible hoses, see figure 4.41.

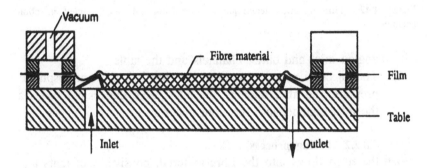

Figure 4.41. Film injection technique: geometry just before opening inlet and outlet valves.

4.8.2.1 The procedure for the production of the flat plate mould

1. Injection and outlet connections are mounted.
2. Over the injection and outlet connections unequally sided, angled aluminium strips are placed to form injection and outlet channels. Slots in one side of the angle strip form the injection and outlet gates, see figure 4.42.
3. The fibre material for the flat plate mould is placed on the table overlapping the angled strips.
4. The film is placed on the table overlapping the rubber strips and the frame on top of the film.
5. The vacuum connection of the closing rim is connected to the vacuum air pump, the injection connection to the resin vessel and the outlet connection to the overflow vessel.
6. With the inlet closed, a vacuum is created in the production cavity (20 mbar), pressing the film onto the fibre material,

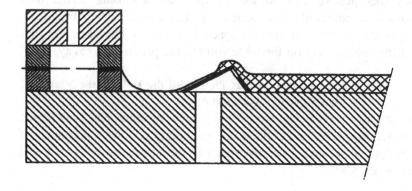

Figure 4.42. Film injection technique detail of unequal sided angled injection channel.

angled inlet and outlet channels and the table.
7. The injection pressure is adjusted to just below the atmospheric pressure and after opening the inlet valve the resin flows via the inlet channel into the fibre material.

4.8.2.2. Filling process

When the resin flows into the fibre material, consisting of mats and woven fabrics, the film is lifted by the combination of injection pressure and compression force in the fibre material. As the injection pressure is below the atmospheric pressure, the film is in contact with the fibre material, see figure 4.43.

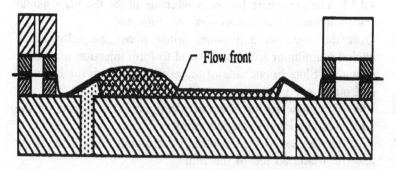

Figure 4.43. Film injection technique when filling is in progress.

As a result of the decompression of the fibre material the permeability increases considerably and consequently the flow front velocity is relatively high. Outside the fibre material the film fits closely on the table and no short-cut channels are formed. When the fibre material is filled completely the fluid pressure is adjusted to the value giving the right thickness of the plate mould. It has to be mentioned that as the film is very flexible, the fibre material is subjected to the same compressing force, at every spot. As the fibre fraction varies over the plate, the distance between film and table varies. Places with less fibre material are compressed more than places with more fibre material. As a result of this the variation in permeability is smaller than in a normal vacuum injection mould, and the flow front less undulated.

In the finished plate mould the thickness of the mould varies but the fibre fraction is more uniform than with the normal vacuum injection technique. One side of the mould is flat and smooth, the side formed by the film is not perfectly flat, but is smooth and glossy. For the production of the second mould part, a more complicated procedure is used.

4.8.2.3 Production of the mould, for the double curved side of the scale model surfboard (Figure 4.44)

Positive models of the surfboard, inlet and outlet channels and grooves for the closing rubbers are placed on the table. The model for the surfboard has to be made by the pattern shop, while the models for channels and grooves are made in the laboratory. In order to be able to use the same closing rubbers for all the moulds to be made in the laboratory, a special mould was made in which silicon rubber positive models can be cast. For the injection outlet channels and the grooves for the closing rubbers the same models are used.

4.8.2.4 The procedure for the production of the mould

1. On the table are placed:
   - Positive model of the scale model surfboard
   - Silicon rubber models of the closing rim
   - Silicon rubber models for inlet and outlet channels
   - Positive models of the inlet and outlet gates
   - The unequal sided angle aluminium strips
   - The fibre material of the mould part
2. The rest of the procedure is similar to the procedure for the flat plate mould described before. It has to be mentioned that:
   - For the closing rim of the mould, a sandwich construction was chosen in order to obtain a rigid closing rim.

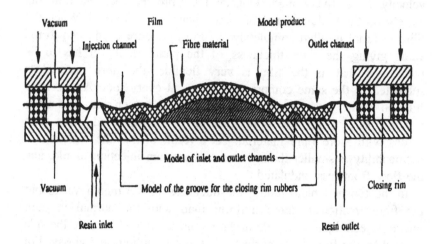

Figure 4.44. Production of the mould part for the scale model surfboard by means of the film vacuum injection technique.

> - Inlet and outlet channels were not finished at this stage, as the computer simulation programme was not available. An annular channel was formed which later, in the first production experiments, was partly plugged by the silicon rubber model to form separate inlet and outlet channels. In this way the correct geometry to obtain voidfree production can easily be determined experimentally.

With the moulds produced by the vacuum film injection technique scale model surfboards were produced, using the normal vacuum injection technique. In order to facilitate placing the fibre material in the moulds preforms can be made.

4.9 PREFORMS

As products usually have a double curved shell-like structure and the fibre material is produced as a flat thin sheet, placing the material in the mould cavity without folds, is difficult.

Preforming the fibre material in a special process shortens the cycle time of the RTM process considerably. Special preform press processes are used when large numbers of preforms have to be produced, using hydraulic presses or special spraying up techniques. In the author's laboratory the vacuum film process is used in the

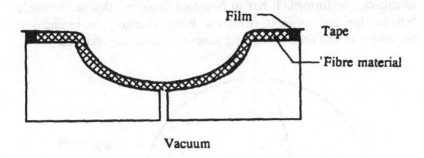

Figure 4.45. Outline of the film vacuum preform process.

RTM mould or a special preform mould, see figure 4.45. A thermoplastic powder is brushed onto the fibre material, the material is preformed, heated and thereupon cooled, in this way stabilising the geometry of the preform. Currently special preformable mats are available which can be used in the preforming processes. When a

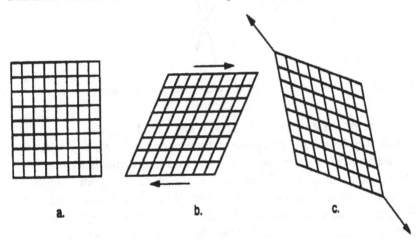

Figure 4.46. Outline of a woven fabric; a) being at rest, b) sheared in two different ways, c) pulled at corners.

preform is used a perfect fit of the fibre material in the mould is achieved, avoiding short-cut channels in the filling process. When woven fabrics are used, the extensive shear deformation of the fabrics, see figure 4.46, helps to avoid folds or pleats. In figure 4.47 it is shown that, when sufficient shear deformation is possible, even a half sphere can be formed. It has to be noted however, that as the angle between the yarn systems (weft and warp) changes, the mechanical properties of the fibre reinforced plastic material are changed.

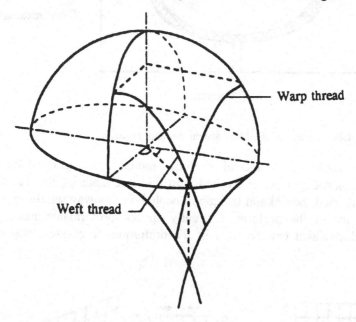

Figure 4.47. Deformation of a woven fabric draped over a sphere.

4.10 VARIANTS OF THE RTM TECHNIQUE

When very large runs of a product have to be produced, e.g., parts for the automotive industry, high pressure injection methods such as very strong and rigid metal mould or Structural Reaction Injection Moulding (SRIM) techniques are used. When high pressure injection is used the fibre material can be compressed by the large difference between liquid pressure and air pressure, see figure 4.48. As a result of the compression of the fibre material a slot is formed between the fibre material and the mould. The velocity of the resin in the slot is very high and the fibre material is filled perpendicular to the sheet of the fibre material. As the thickness of those shell-like structures is small, filling of the mould is completed in a very short time.

To achieve very short cycle times for the production process very reactive resin systems have to be used. As class A surface quality is essential for the automotive industry the effects of reaction shrinkage have to be compensated.

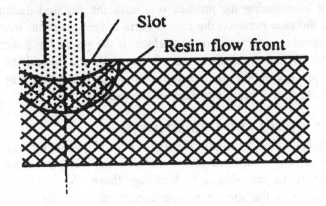

Figure 4.48. Slot and sperical resin flow front at the central inlet gate.

4.11 EFFECTS OF REACTION SHRINKAGE OF THE RESIN ON SURFACE QUALITY OF PRODUCTS

In the curing process of most types of resins, reaction shrinkage will occur. When a liquid resin is cured in a simple open mould, it appears that the free surface is not perfect. In the free surface of the resin in the mould, grooves (called lakes) are formed as shown in figure 4.49.

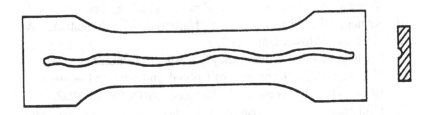

Figure 4.49. Formation of a groove in the free surface of a casted dogbone test bar.

When a flexible mould is used, as e.g., in vacuum injection techniques, the mould surface presses on the resin in the mould compensating local shrink differences and resulting in a perfectly smooth surface of the product. When rigid moulds are used the moulds are generally closed by hydraulic presses. When the mould is filled, the distance between the mould parts has to be adjusted in such a way that after curing the product will have the required thickness. When the distance between the mould parts is kept constant, more or less free shrinkage occurs and the surface of the product is generally not perfect. When a distinct load is kept on the mould by the hydraulic press, compensation of the local shrinkage differences will occur, resulting in a perfect surface of the product.

4.12 CONCLUDING COMMENTS
As motorboats, sailing yachts and even smaller ships are, for a greater part, made of fibre reinforced plastic material, the RTM technique can be of interest to the shipyards building those ships. Using this technique reduces the styrene vapour concentration in the air of the shipyards and a voidfree hull with a high fibre-volume fraction can be produced. As the vacuum film injection technique uses only one mould that is hardly more expensive than the hand lay-up mould, this technique can be of interest for the production of high quality hulls.

4.13 REFERENCES
1] Lubin, G., (ed.), "Handbook of Composites", Van Nostrand Reinhold, New York, 1982.
2] Greenkorn, R.A., "Flow Phenomena in Porous Media", Dekker, New York, 1983.
3] Verruijt, A., "Ground Water Flow", McMillan, London, 1982.
4] Kronig, R., "Leerboek der Natuurkunde", Scheltema & Holkema, Amsterdam, 1962.
5] Chatterjee, P.K., "Absorbency", Elsevier, Amsterdam, 1985.
6] Hiemenz, P.C., "Principles of Colloid and Surface Chemistry", Books Demand, New York, Undated, ISBN 0835735052.
7] Hildeband, J.H., Prausnitz, J.H., Scott, R.L., "Regular and Related Solutions", Kriger, New York, 1970.
8] Shinoda, K., "Principles of Solution and Solubility", Books Demand, New York, 1978.
9] Brandrup, J., Immergut, E.H., "Polymer Handbook", Wiley, New York, 1989.
10] Crank, J., "The Mathematics of Diffusion", OUP, Oxford, 1975.

5 AN ENGINEERING APPROACH TO THE PREDICTION OF ELASTIC PROPERTIES OF A LAMINATE

5.1 INTRODUCTION

Early approaches to calculate mechanical properties of fibre reinforced composites were based on simple models of fibre-matrix interaction [1] or on rule-of-mixtures formulae [2] in which fibres and matrix were assumed to act together in series or parallel. Application of energy theorems to composite behaviour showed that the assumptions of parallel or series connection between fibres and matrix yield upper and lower bounds on effective moduli. Subsequent work [3-9] dealt primarily with the requirement of energy methods to achieve closer spacing of the bounds and to provide solutions for a range of possible fibre-matrix geometries. The purpose of this Chapter is to provide the reader with a simple exposition of procedures for calculating elastic properties of composites which are amenable to easy, practical application.

5.2 VOLUME AND WEIGHT FRACTIONS

One of the primary factors which determines the properties of composites is the relative proportions of the fibres and matrix. The relative proportions can be in terms of weight or volume fractions. Weight fractions are easier to obtain during fabrication or by an experimental method after fabrication. Volume fractions, on the other hand, are more convenient for theoretical calculations. Hence it is desirable to determine expressions for conversion between weight and volume fractions.

Consider a case with weights/volumes of the fibres, matrix and net composite being w_f/v_f, w_m/v_m and w_c/v_c respectively. (Subscripts f, m and c from here on are consistently used to represent fibres, matrix and composite.) Let the volume and weight fractions be denoted by V and W respectively. These are defined as follows:

$$v_c = v_f + v_m \tag{5.1a}$$

$$V_f = \frac{v_f}{v_c} \quad , \quad V_m = \frac{v_m}{v_c} \tag{5.1b}$$

and

$$w_c = w_f + w_m \tag{5.1c}$$

$$W_f = \frac{w_f}{w_c} \quad , \quad W_m = \frac{w_m}{w_c}. \tag{5.1d}$$

To establish conversion relationships between the weight and volume fractions, it is necessary to obtain the density of the composite, ρ_c. This can be obtained by initially re-writing equation (5.1c) as follows:

$$\rho_c v_c = \rho_f v_f + \rho_m v_m. \tag{5.2}$$

Dividing both sides by v_c and substituting the definitions of volume fractions from equation (5.1b) yields

$$\rho_c = \rho_f V_f + \rho_m V_m. \tag{5.3}$$

By similar manipulations in equation (5.1a), the density of the composite material in terms of weight fractions can also be shown to be:

$$\rho_c = \frac{1}{(W_f/\rho_f) + (W_m/\rho_m)}. \tag{5.4}$$

Now the conversion between the weight and volume fractions can be determined by considering the definition of the weight fraction and replacing in it the weights by the products of density and volume as follows:

$$W_f = \frac{w_f}{w_c} = \frac{\rho_f\, v_f}{\rho_c\, v_c}$$

or

$$W_f = \frac{\rho_f}{\rho_c} V_f \quad : \quad W_m = \frac{\rho_m}{\rho_c} V_m. \tag{5.5}$$

Equations (5.3-5.5) have been derived for a composite material with only two constituents but can be generalised for an arbitrary number of constituents. The generalised equations are:

$$\rho_c = \sum_{i=1}^{n} (\rho_i V_i) \; : \; \rho_c = \frac{1}{\sum_{i=1}^{n} (W_i/\rho_i)} \; : \; W_i = \frac{\rho_i}{\rho_c} V_i . \qquad (5.6)$$

It must be mentioned here that the theoretically calculated composite density is not always in agreement with experimentally derived values. This is because of the presence of voids or air bubbles. If the theoretical and experimentally determined composite densities are ρ_{ct} and ρ_{ce} respectively, then the volume fraction of voids, V_v, is given by:

$$V_v = \frac{\rho_{ct} - \rho_{ce}}{\rho_{ct}} . \qquad (5.7)$$

The influence of voids on mechanical properties is discussed in section 5.8.3 below.

5.3 LONGITUDINAL STRENGTH AND STIFFNESS

A unidirectional composite may be modelled by assuming fibres to be uniform in properties and diameter, continuous and parallel throughout. If it is further assumed that a perfect bond exists between the fibres and matrix and no slippage occurs at the interface, then the longitudinal strains experienced by the composite, fibres and matrix are equal:

$$\varepsilon_{cL} = \varepsilon_f = \varepsilon_m = \varepsilon \qquad (5.8)$$

the load carried by the composite, P_{cL} , is shared between the fibres, P_f, and matrix, P_m:

$$P_{cL} = P_f + P_m \qquad (5.9)$$

The loads, in turn, can be written in terms of the stresses and corresponding cross-sectional areas. Thus:

$$P_{cL} = \sigma_{cL} A_c = \sigma_f A_f + \sigma_m A_m$$

or

$$\sigma_{cL} = \sigma_f \frac{A_f}{A_c} + \sigma_m \frac{A_m}{A_c} . \qquad (5.10)$$

However, for composites with parallel fibres, the volume fractions are equal to the area fractions such that:

$$V_f = \frac{A_f}{A_c} \; , \quad V_m = \frac{A_m}{A_c}. \tag{5.11}$$

Thus, the composite stress along the fibre direction, σ_{cL} is:

$$\sigma_{cL} = \sigma_f V_f + \sigma_m V_m. \tag{5.12}$$

Equation (5.12) can be differentiated with respect to strain yielding:

$$\frac{d\sigma_{cL}}{d\varepsilon} = \frac{d\sigma_f}{d\varepsilon} V_f + \frac{d\sigma_m}{d\varepsilon} V_m$$

where $(d\sigma/d\varepsilon)$ represents the slope of the stress-strain curve or the corresponding elastic modulus. Thus, the above equation can be re-written, in terms of the composite longitudinal modulus, as:

$$E_{cL} = E_f V_f + E_m V_m. \tag{5.13}$$

Equations (5.12) and (5.13) are termed rule-of-mixtures formulations. A slightly modified version suggested by Whitney and Riley [10] takes account of variations of fibre alignment or straightness by introducing a factor, K, (with $0 \le K \le 1$) such that:

$$E_{cL} = KE_f - (KE_f - E_m) V_m. \tag{5.14}$$

With K = 1, the above equation reverts to its classic form in equation (5.13). This equation is preferred to Tsai's proposal [11] of $E_{cL} = K[E_f -(E_f - E_m)V_m]$ as this latter has been found in some cases to cause singularity of the compliance matrix and to give negative laminate moduli for low values of K [12]. The rule-of-mixtures value is theoretically exact if the fibre and matrix Poisson's ratios, υ_f and υ_m, are equal. Equations (5.12) and (5.13) can therefore be generalised for multi-phase materials as:

$$\sigma_{cL} = \sum_{i=1}^{n} \sigma_i V_i \quad : \quad E_{cL} = \sum_{i=1}^{n} E_i V_i. \tag{5.15}$$

5.4 TRANSVERSE STIFFNESS
As in the previous section, the fibres are assumed to be uniform in properties and diameter, continuous and parallel throughout the composite. The composite, made up of layers representing fibres and matrix material, is stressed in a transverse direction (perpendicular to the parallel fibres) as shown in figure 5.1. Each layer has the same area on which the load acts. It is clear that each layer will carry the

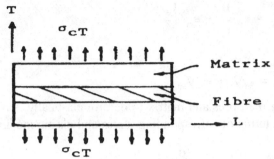

T

σ_{cT}

Matrix

Fibre

L

σ_{cT}

a)   Application of uniform stress σ_{cT}

T

Deformed
Shape

L

b)   Deformation due to σ_{cT}

Figure 5.1. Representative element under transverse normal stress, σ_{cT}.

same load and experience the same stress:

$$\sigma_{cT} = \sigma_f = \sigma_m. \tag{5.16}$$

Because each layer is assumed to be uniform in thickness, it follows that the cumulative thicknesses of the fibre and matrix layers will be proportional to their respective volume fractions. In this case, the composite transverse elongation, δ_{cT}, is the sum of the fibre and matrix elongations, δ_f and δ_m respectively:

$$\delta_{cT} = \delta_f + \delta_m. \tag{5.17}$$

The elongation of the material can be written as the product of the strain and its cumulative thickness:

$$\delta_{cT} = \varepsilon_{cT} t_c, \quad \delta_f = \varepsilon_f t_f, \quad \delta_m = \varepsilon_m t_m. \tag{5.18}$$

Substituting from equation (5.18) into equation (5.17) gives:

$$\varepsilon_{cT} t_c = \varepsilon_f t_f + \varepsilon_m t_m. \tag{5.19}$$

Dividing both sides of equation (5.19) by t_c and recognising that thickness is proportional to volume fraction:

$$\varepsilon_{cT} = \varepsilon_f \frac{t_f}{t_c} + \varepsilon_m \frac{t_m}{t_c}$$

or

$$\varepsilon_{cT} = \varepsilon_f V_f + \varepsilon_m V_m. \tag{5.20}$$

Assuming the fibres and matrix to deform elastically, strain can be re-written in terms of corresponding stress and elastic moduli as follows:

$$\frac{\sigma_{cT}}{E_c} = \frac{\sigma_f}{E_f} V_f + \frac{\sigma_m}{E_m} V_m.$$

Noting, from equation (5.16) that the stresses in the composite, fibre and matrix are equal, the above expression can be simplified to:

$$\frac{1}{E_{cT}} = \frac{V_f}{E_f} + \frac{V_m}{E_m}. \tag{5.21}$$

For a composite made from n different materials, the above equation can be generalised as:

$$E_{cT} = \frac{1}{\sum_{i=1}^{n} (V_i / E_i)}. \tag{5.22}$$

The manner in which the transverse modulus, E_{cT}, depends on the volume of fibres is shown in figure 5.2. It can be noticed that for a volume of fibres as large as 50%, the effect of fibres only increases the modulus to approximately twice the matrix modulus, even though fibre modulus is taken as 30 times the matrix modulus. It is not until much higher volumes of fibres are present that the fibres themselves have any substantial effect in raising the value of the transverse modulus. This is in sharp contrast to the effect of fibres on the longitudinal modulus, E_{cL}.

5.5 PREDICTION OF SHEAR MODULUS

The derivation of the in-plane shear modulus is based on the same model as that used for determining transverse modulus in the previous section. The representative element shown in figure 5.1 is considered to be subjected to uniform, shear and complementary shear stresses along the boundaries. Shearing stress on the fibres and matrix are equal. Thus:

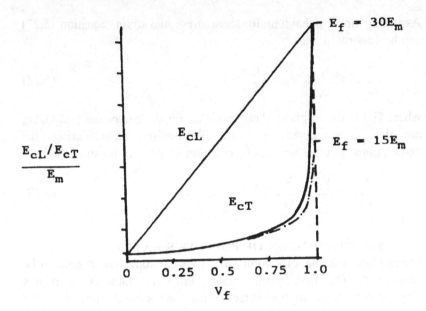

Figure 5.2. Variation of E_{cL} and E_{cT} with fibre content.

$$\tau_c = \tau_f + \tau_m. \tag{5.23}$$

The total shear deformation of the composite, Δ_c, is the sum of the shear deformations of the fibre and matrix, Δ_f and Δ_m:

$$\Delta_c = \Delta_f + \Delta_m. \tag{5.24}$$

The shear deformation in each constituent can be expressed as the product of the shear strain and the corresponding thickness:

$$\Delta_c = \gamma_c t_c, \quad \Delta_f = \gamma_f t_f, \quad \Delta_m = \gamma t_m. \tag{5.25}$$

Substitution of equation (5.25) in equation (5.24) yields:

$$\gamma_c t_c = \gamma_f t_f + \gamma_m t_m. \tag{5.26}$$

Dividing both sides by the thickness t_c and recognising that thickness is proportional to volume fraction:

$$\gamma_c = \gamma_f \frac{t_f}{t_c} + \gamma_m \frac{t_m}{t_c}$$

or

$$\gamma_c = \gamma_f V_f + \gamma_m V_m. \tag{5.27}$$

Assuming linear behaviour for shear stress and strain, equation (5.27) can be re-written as:

$$\frac{\tau_c}{G_{LT}} = \frac{\tau_f}{G_f} \cdot V_f + \frac{\tau_m}{G_m} \cdot V_m \qquad (5.28)$$

where G_{LT} is the in-plane shear modulus for the composite and G_f/G_m are the shear moduli for the fibres-matrix. Considering the assumptions of equation (5.23), equation (5.28) can be simplified as:

$$\frac{1}{G_{LT}} = \frac{V_f}{G_f} + \frac{V_m}{G_m} . \qquad (5.29)$$

5.6 PREDICTION OF POISSON'S RATIO

For in-plane loading of a composite, two Poisson's ratios need to be considered. The first, termed major Poisson's ratio (υ_{LT}), relates longitudinal stress to transverse strain. The second, termed minor Poisson's ratio (υ_{TL}), relates the transverse stress to longitudinal strain.

The major Poisson's ratio can be determined by using the same model as that used for predicting the transverse modulus. However, the load is applied parallel to the (fibres or) layers in the model. The deformation pattern (as dotted lines) is shown, superimposed on the un-stressed fibre-matrix element, in figure 5.3.

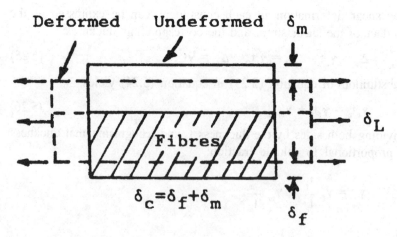

Figure 5.3. Model for prediction of Poisson's ratio.

Transverse strains in the composite fibres and matrix can be written in terms of longitudinal strains and Poisson's ratios as follows:

$$(\varepsilon_T)_c = -\upsilon_{LT}(\varepsilon_L)_c, \quad (\varepsilon_T)_f = -\upsilon_f(\varepsilon_L)_f, \quad (\varepsilon_T)_m = -\upsilon_m(\varepsilon_L)_m, \quad (5.30)$$

where υ_f/υ_m are Poisson ratios for the fibre and matrix respectively. Transverse deformations can now be written as the product of strain and the corresponding thickness:

$$\delta_c = -t_c\,\upsilon_{LT}(\varepsilon_L)_c, \quad \delta_f = -t_f\,\upsilon_f(\varepsilon_L)_f, \quad \delta_m = -t_m\cdot\upsilon_m(\varepsilon_L)_m. \quad (5.31)$$

The total deformation of the composite is composed of the fibre and matrix contributions.

$$-t_c\,\upsilon_{LT}(\varepsilon_L)_c = -t_f\,\upsilon_f(\varepsilon_L)_f - t_m\,\upsilon_m(\varepsilon_L)_m. \quad (5.32)$$

Because the longitudinal strains in the composite, fibre and matrix, due to the longitudinal stress, are equal, equation (5.32) becomes:

$$t_c\,\upsilon_{LT} = t_f\upsilon_f + t_m\upsilon_m. \quad (5.33)$$

Dividing both sides of equation (5.33) by t_c and noting again that thickness is proportional to the volume fraction:

$$\upsilon_{LT} = \upsilon_f V_f + \upsilon_m V_m. \quad (5.34)$$

This is the rule-of-mixtures formulation for the major Poisson's ratio. The minor Poisson's ratio can be derived from the classic equation:

$$\frac{\nu_{LT}}{E_L} = \frac{\nu_{TL}}{E_T}. \quad (5.35)$$

5.7 ALTERNATIVE APPROACHES TO PREDICTION OF CONSTANTS

5.7.1 Elasticity Methods for Stiffness Prediction

The simple models described in the previous sections are not mathematically rigorous. Because, in a real composite, parallel fibres and matrix are dispersed in a random fashion it follows that the likelihood is one of load sharing between fibres and matrix. Hence the assumption of equal stresses in the fibre and matrix for transverse loading, for example, is invalid. The equality of stresses also results in a mismatch of strains at the fibre-matrix interface. A further inaccuracy arises because of mismatch of Poisson's ratios of the fibres and matrix which induces loads in the fibres and matrix perpendicular to the load with no net resultant force on the composite in that direction. A mathematically rigorous solution with a full compatibility of displacements across the fibre-matrix boundary is accomplished through the use of the theory of elasticity. Such methods can be divided into three categories.

Bounding Techniques: Energy theorems of classical elasticity are used to obtain bounds on the elastic properties [3,6] with the minimum complementary energy theorem yielding the lower bound and the minimum potential energy theorem yielding the upper bound.

Exact Solutions: This consists of assuming the fibres to be arranged in regular, periodic arrays. There are few closed form solutions using classical techniques because of difficulties arising from complex geometries of the reinforcement. Solution techniques include series development, use of complex variables, finite difference and finite element analyses [13,14].

Self Consistent Model: Here a single fibre is assumed to be embedded in a concentric cylinder of matrix material. This, in turn, is embedded in a homogenous material that is macroscopically the same as the composite being studied. This method [9,10] has the advantage that its results are applicable to regular or irregular packing of fibres.

A large amount of data has been generated through the use of such procedures. The results are in the form of curves or complicated equations; their adaptability to design procedures is limited. For design purposes, the requirement is to have simple expressions amenable to quick computation, even though the estimates may be approximate.

5.7.2 Halpin-Tsai Equations for Transverse Modulus

Semi-empirical equations derived by Halpin and Tsai [15] corresponding to a generalisation of accurate elasticity solutions for microscopic fibre-matrix models, provide a simpler means of estimating moduli. These appear to be sufficiently accurate for most composites except, possibly, those with a high fibre content, $V_f > 0.6$ [16]. The Halpin-Tsai equation for transverse modulus can be written as:

$$\frac{E_{cT}}{E_m} = \frac{1+\xi\eta V_f}{1-\eta V_f} .$$ (5.36)

where:

$$\eta = \frac{(E_f/E_m)-1}{(E_f/E_m)+\xi}$$

in which ξ is a measure of reinforcement and is dependent on fibre/packing geometry and loading conditions; it is determined empirically through comparisons with the results from elasticity solutions. For the usual case of circular-section fibres, satisfactory results are obtained by taking $\xi = 2$.

5.7.3 Halpin-Tsai Equations for Shear Modulus

The rule-of-mixtures model for predicting shear modulus of a unidirectional composite suffers from the same limitations as mentioned above. Hence, recourse can again be made of the Halpin-Tsai equations, which, in case of the shear modulus, can be written as:

$$\frac{G_{LT}}{G_m} = \frac{1+\xi\eta V_f}{1-\eta V_f} .$$ (5.37)

where:

$$\eta = \frac{(G_f/G_m)-1}{(G_f/G_m)+\xi}$$

Halpin and Tsai have suggested that $\xi = 1$. The way in which shear modulus varies with the volume of fibres is shown in figure 5.4. It can be noticed that, as for the transverse modulus, it is the shear modulus of the matrix that determines the value of the shear modulus of the composite (unless the fibre-volume becomes excessively high).

5.8 PRACTICAL CONSIDERATIONS

5.8.1 Moduli of Short-Fibre Composites

For a composite reinforced by aligned (unidirectional) short-fibres, it has been shown [17] that longitudinal modulus, E_{cL}, may be estimated using equation (5.36) with $E_{cT} = E_{cL}$ and $\xi = 2l/d$, where l is fibre length and d is fibre diameter. It is evident from this that if $l/d > 100$, that is $l > 1.0\,mm$ for a typical fibre diameter of $0.01\,mm$, then E_{cL}

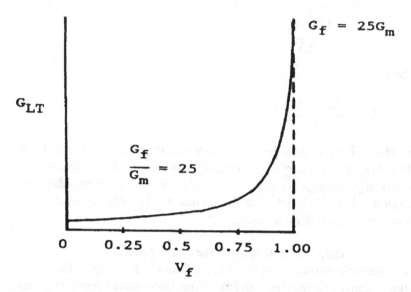

Figure 5.4. Variation of G_{LT} with fibre content.

tends towards the rule-of-mixture formulation given in equation (5.13).

Chopped strand mat (CSM) laminates used in marine construction generally have random fibre orientation with isotropic in-plane moduli which may be derived from the moduli of an aligned short-fibre composite [18] using the following empirical equations

$$E = \frac{3}{8} E_{cL} + \frac{5}{8} E_{cT} . \tag{5.38}$$

$$G = \frac{1}{8} E_{cL} + \frac{1}{4} E_{cT} . \tag{5.39}$$

where E_{cL} and E_{cT} are derived from equations (5.13) and (5.36) respectively. The corresponding Poisson's ratio follows from $\upsilon = (E/2G)-1$. The moduli of a composite with randomly oriented short (or continuous) fibres can also be estimated [19] from the following expressions:

$$E = (U_1 + U_4) (U_1 - U_4)/U_1 \tag{5.40}$$

$$G = (U_1 - U_4)/2 \tag{5.41}$$

$$\upsilon = U_4/U_1 \tag{5.42}$$

where U_1 and U_4 are invariants defined as [20]:

$$U_1 = (3C_{11} + 3C_{22} + 2C_{12} + 4C_{66})/8 \qquad (5.43)$$

$$U_4 = (C_{11} + C_{22} + 6C_{12} - 4C_{66})/8 \qquad (5.44)$$

and

$$C_{11} = E_{cL}/(1 - \upsilon_{LT}.\upsilon_{TL}) \qquad (5.45)$$

$$C_{22} = E_{cT}/(1 - \upsilon_{LT}.\upsilon_{TL}) \qquad (5.46)$$

$$C_{12} = \upsilon_{LT}.E_{cT}/(1 - \upsilon_{LT}.\upsilon_{TL}) \qquad (5.47)$$

$$C_{66} = G_{LT} \qquad (5.48)$$

Sprayed chopped strand laminate may also be assumed to have random fibre direction, though it should be noted that the spraying process may give rise to some bias in fibre direction. A more refined method of assessing the stiffness of short-fibre composites, outlined by Cervenka [21], accounts statistically for variations in fibre length and orientation.

5.8.2 Treatment of Cloth Laminates
In order to simplify the analysis of woven cloth laminates, avoiding the need to treat such materials as a large number of overlaid bidirectional plies, a notion of "ply efficiency" can be introduced. A bidirectional cloth ply may then be treated as a pair of equivalent superimposed, orthogonally oriented, unidirectional plies, each occupying the same space as the actual ply [20]. The equivalent unidirectional plies of a balanced cloth material, for example, would have an efficiency of 0.5. For a biased fabric, unequal fibre distribution may be represented by adopting different efficiencies in the orthogonal directions. For example, in a material with 3 to 1 fibre distribution, the constituent plies would have efficiencies of 0.75 in the dominant fibre direction and 0.25 in the secondary direction.

5.8.3 Influence of Voids
As mentioned in section 5.2 above, in addition to fibres and matrix, a composite may contain voids in the form of air bubbles. Assuming these bubbles to be spherical in shape, and are uniformly distributed throughout the material, accurate evaluation of moduli may be using results obtained by Hashin [4]. Alternatively and more simply, if the

void content is small and contained entirely within the matrix, voids can be represented adequately by adopting a corrected matrix modulus, E_m, and Poisson ratio, v_m, based on rule-of-mixture assumptions. Assuming the void content to be V_v, and noting that the relationship $V_f + V_m + V_v = 1$ holds, corrected matrix properties are:

$$\overline{E_m} = \frac{E_m V_m}{\overline{V_m}}.$$ (5.49)

$$\overline{v_m} = \frac{v_m V_m}{\overline{V_m}}.$$ (5.50)

5.9 REFERENCES

1] Shaffer, B.W., "Stress-strain Relations of Reinforced Plastics Parallel and Normal to their Internal Filaments", AIAA J., **2** (2), February 1984. pp 348-352.

2] Sonneborn, R.H., "Fibreglass Reinforced Plastics", Van Nostrand Reinhold, New York, 1966.

3] Paul, B., "Predictions of Elastic Constants of Multiphase Materials", Trans. Met. Soc. AIME, **218**, February 1968. pp 36-41.

4] Hashin, Z., "The Elastic Moduli of Heterogeneous Materials", J. App. Mech., **29** (1), March 1962. pp 143-150.

5] Hashin, Z., Shtrikman, S., "Note on a Variational Approach to the Theory of Composite Elastic Materials", J. Franklin Inst., **271**, April 1961. pp 336-341.

6] Hashin, Z., Rosen, B.W., "The Elastic Moduli of Fiber Reinforced Materials", J. App. Mech., **31** (2), June 1964. pp 223-232.

7] Hashin, Z., "Theory of Mechanical Behaviour of Heterogeous Media", App. Mech. Rev., **17** (1), January 1964. pp 1-9.

8] Hill, R., "Elastic Properties of Reinforced Solids: Some Theoretical Principles", J. Mech. Phy. Solids, **11** (5), September 1963. pp 357-372.

9] Hill, R., "Theory of Mechanical Properties of Fibre-Strengthened Materials", J. Mech. Phy. Solids, **12** (4), September 1964. pp 199-218.

10] Whitney, J.M., Riley, M.B., "Elastic Properties of Fiber Reinforced Composite Materials", AIAA J., **4**, (9), September

1966. pp 1537-1542.

11] Tsai, S.W., "Structural Behaviour of Composite Materials", NASA CR-71, July 1964.

12] Smith, C.W., "Calculation of Elastic Properties of GRP Laminates for Use in Ship Design", Proc. Symp. *GRP Ship Construction*, RINA, London, October 1972. pp 69-80.

13] Agarwal, B.D., Broutman, L.J., "Analysis and Performance of Fibre Composites", John Wiley and Sons, New York, 1990.

14] Adams, D.F., Doner, D.R., "Transverse Normal Loading of a Unidirectional Composite", J. Comp. Mater., 1 (2), April 1967. pp 152-166.

15] Halpin, J.C., Tsai, S.W., "Effects of Environmental Factors on Composite Materials", AFML-TR 67-423, June 1969.

16] Jones, R.M., "Mechanics of Composite Materials", McGraw Hill, New York, 1975.

17] Halpin, J.C., "Stiffness and Expansion Estimates for Oriented Short Fibre Composites", J. Comp. Mater., 3 (3), October 1969. pp 732-734.

18] Hull, D., "An Introduction to Composite Materials", CUP, Cambridge, 1981.

19] Manera, M., "Elastic Properties of Randomly Oriented Short Fiber-Glass Composites", J. Comp. Mater., 11 (2), April 1977. pp 235-247.

20] Smith, C.S., "Design of Marine Structures in Composite Materials", Elsevier Applied Science, London, 1990.

21] Cervenka, A., "Stiffness of Short-Fibre Reinforced Materials", Proc. 6th Intl. Conf. *Composite Materials*, London, 1987.

6 MECHANICS OF ORTHOTROPIC LAMINAE

6.1 INTRODUCTION

Composite materials like fibre reinforced plastics (FRP) are heterogeneous and anisotropic materials. If the fibres are distributed uniformly in the matrix material, in terms of global stiffness computations FRP can be understood as being statistically homogeneous. Then stresses and strains are volumetrically averaged. There is no attention given to local differences in stresses and strains of fibres and matrix. In that case the mechanics of homogeneous, anisotropic materials can be used to characterise the stiffness behaviour of FRP's.

To indicate the position of a point an orthogonal coordinate system with axes x_1, x_2 and x_3 is introduced. The three-dimensional stress is represented by the stress tensor or stress matrix $[\sigma]$:

$$[\sigma] = \begin{bmatrix} \sigma_{11} & \sigma_{12} & \sigma_{13} \\ \sigma_{21} & \sigma_{22} & \sigma_{23} \\ \sigma_{31} & \sigma_{32} & \sigma_{33} \end{bmatrix} ; \quad \sigma_{ji} = \sigma_{ij}, \text{ with } i,j \in \{1,2,3\}, \tag{6.1a}$$

where σ_{11}, σ_{22} and σ_{33} are normal stress components and the remaining elements of the matrix are shear stress components. Due to equilibrium the strain matrix is symmetric with respect to the main diagonal as expressed in the right equation of (6.1a).

In contracted notation the stresses are represented by the stress vector $\{\sigma\}$:

$$\{\sigma^T\} = [\sigma_1 \ \sigma_2 \ \sigma_3 \ \sigma_4 \ \sigma_5 \ \sigma_6], \tag{6.1b}$$

where $\sigma_1 = \sigma_{11}$, $\sigma_2 = \sigma_{22}$, $\sigma_3 = \sigma_{33}$, $\sigma_4 = \sigma_{23}$, $\sigma_5 = \sigma_{31}$, $\sigma_6 = \sigma_{12}$.

(See figure 6.1). The deformations are described by the strain tensor or strain matrix $[\varepsilon]$:

$$[\varepsilon] = \begin{bmatrix} \varepsilon_{11} & \varepsilon_{12} & \varepsilon_{13} \\ \varepsilon_{21} & \varepsilon_{22} & \varepsilon_{23} \\ \varepsilon_{31} & \varepsilon_{32} & \varepsilon_{33} \end{bmatrix} ; \quad \varepsilon_{ji} = \varepsilon_{ij}, \text{ with } i,j \in \{1,2,3\}, \tag{6.2a}$$

where ε_{11}, ε_{22}, and ε_{33} are direct strain components and the remaining elements of $[\varepsilon]$ are shear strain components.

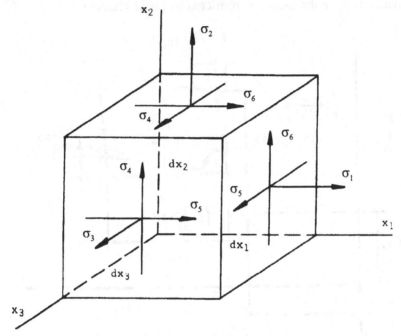

Figure 6.1. Notation and convention for positive direction of the stress components.

Both components ε_{ij} and ε_{ji} represent the same change in angle between x_i and x_j-axis. In contracted notation the strains are represented by the strain vector $\{\varepsilon\}$:

$$\{\varepsilon^T\} = [\varepsilon_1 \ \varepsilon_2 \ \varepsilon_3 \ \varepsilon_4 \ \varepsilon_5 \ \varepsilon_6], \tag{6.2b}$$

where $\varepsilon_1 = \varepsilon_{11}$, $\varepsilon_2 = \varepsilon_{22}$, $\varepsilon_3 = \varepsilon_{33}$, $\varepsilon_4 = 2\varepsilon_{23}$, $\varepsilon_5 = 2\varepsilon_{31}$, $\varepsilon_6 = 2\varepsilon_{12}$.

Notice that in the contracted notation used here the so-called engineering shear strain components are twice the tensor shear strain components.

If the deformations and the displacement gradients are small, the strain components are linearly connected with the gradients of the displacements u_i in x_i-direction (figure 6.2):

$$\varepsilon_1 = \frac{\partial u_1}{\partial x_1} \; ; \; \varepsilon_2 = \frac{\partial u_2}{\partial x_2} \; ; \; \varepsilon_3 = \frac{\partial u_3}{\partial x_3} \; ;$$

$$\varepsilon_4 = \frac{\partial u_2}{\partial x_3} + \frac{\partial u_3}{\partial x_2} \; ; \; \varepsilon_5 = \frac{\partial u_3}{\partial x_1} + \frac{\partial u_1}{\partial x_3} \; ; \; \varepsilon_6 = \frac{\partial u_1}{\partial x_2} + \frac{\partial u_2}{\partial x_1} \; .$$

(6.3)

The stress-strain relations are dependent on the properties of the material. Here the theory is restricted to linear elasticity.

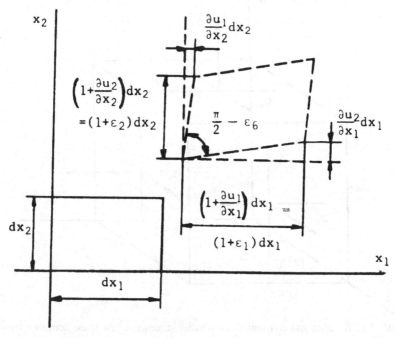

Figure 6.2. The in-plane strains of the x_1x_2-plane in the notation used here.

FRP's are mostly shaped of layered construction, a laminate, composed of laminae with a plane fibre reinforcement such as a unidirectional layer of fibre bundles, a random chopped strand mat or a woven fabric. These FRP layers show material symmetry. Three orthogonal planes of symmetry can be distinguished. So the anisotropy is reduced to orthotropy. For that reason the theory is further restricted to the mechanics of orthotropic laminae and of laminates built of such laminae.

The classical laminate theory (section 7.2) is based on the assumption of plane stress conditions in the different laminae. Then the number of stress components is reduced to the three in-plane components, σ_1, σ_2, and σ_6.

6.2 GENERALISED HOOKE'S LAW

Common reinforcing fibres - glass, carbon and aramid - show approximately linear elasticity up to fracture. If the relatively stiff fibres (the modulus of elasticity of glass is roughly 20 times the modulus of a thermosetting matrix) determine the stiffness of the FRP, it is reasonable to suppose linear elasticity for the whole material. This is especially the case in the fibre direction of a unidirectional FRP (the so-called longitudinal direction). In the transverse direction the stress-strain relation will deviate from linearity. But if the FRP object should be loaded in a transverse direction, it will be necessary to add fibres lying more or less in that direction. If these added fibre layers dominate the elastic behaviour in the former transverse direction, the assumption of linearity is still acceptable.

The empirical Hooke's law supposes that the strain components are proportional to the stress components. In generalised form it results in 6 linear equations:

$$\varepsilon_i = \sum_{j=1}^{6} S_{ij}\, \sigma_j \ . \tag{6.4a}$$

Written in matrix algebra:

$$\begin{bmatrix} \varepsilon_1 \\ \varepsilon_2 \\ \varepsilon_3 \\ \varepsilon_4 \\ \varepsilon_5 \\ \varepsilon_6 \end{bmatrix} = \begin{bmatrix} S_{11} & S_{12} & S_{13} & S_{14} & S_{15} & S_{16} \\ S_{21} & S_{22} & S_{23} & S_{24} & S_{25} & S_{26} \\ S_{31} & S_{32} & S_{33} & S_{34} & S_{35} & S_{36} \\ S_{41} & S_{42} & S_{43} & S_{44} & S_{45} & S_{46} \\ S_{51} & S_{52} & S_{53} & S_{54} & S_{55} & S_{56} \\ S_{61} & S_{62} & S_{63} & S_{64} & S_{65} & S_{66} \end{bmatrix} \begin{bmatrix} \sigma_1 \\ \sigma_2 \\ \sigma_3 \\ \sigma_4 \\ \sigma_5 \\ \sigma_6 \end{bmatrix} , \tag{6.4b}$$

or in short:

$$\{\varepsilon\} = [S]\,\{\sigma\} \ . \tag{6.4c}$$

The coefficients S_{ij} - the so-called compliances - constitute a 6x6 compliance matrix [S]. The stress components are also linear functions of the strain components:

$$\{\sigma\} = [S]^{-1}\{\varepsilon\} = [C]\{\varepsilon\} \ . \tag{6.5}$$

The stiffness matrix [C] is equal to the inverse of the compliance matrix, and the elements C_{ij} are the stiffness constants of the material, with the dimension of stress.

The determinant of the square matrix [S] has the symbol $|S|$. The determinant of the submatrix of [S] developed by deleting the ith row and the jth column is symbolised by $|S_{(ij)}|$. Using these symbols the well-known Cramer's rule can be written as:

$$C_{ij} = (-1)^{i+j} \frac{|S_{(ji)}|}{|S|} .$$
(6.6)

It is important to recognise the order of the subscripts here.

For thermodynamic reasons there are 15 relations between the stress components and the strain components, well-known as the reciprocal equations:

$$\frac{\partial \sigma_i}{\partial \varepsilon_j} = \frac{\partial \sigma_j}{\partial \varepsilon_i} .$$
(6.7)

A consequence of this reciprocity is the symmetry of the matrices [C] and [S] with respect to the main diagonal:

$$C_{ji} = C_{ij} ; \quad S_{ji} = S_{ij} .$$
(6.8)

That is why generally 21 independent elastic constants (6x6 = 36 compliances or stiffness constants minus 15 relations among them) determine the linear elastic behaviour of an arbitrary anisotropic material. And so:

$$
\begin{bmatrix} \sigma_1 \\ \sigma_2 \\ \sigma_3 \\ \sigma_4 \\ \sigma_5 \\ \sigma_6 \end{bmatrix} = \begin{bmatrix} C_{11} & C_{12} & C_{13} & C_{14} & C_{15} & C_{16} \\ & C_{22} & C_{23} & C_{24} & C_{25} & C_{26} \\ & & C_{33} & C_{34} & C_{35} & C_{36} \\ & & & C_{44} & C_{45} & C_{46} \\ \text{symm.} & & & & C_{55} & C_{56} \\ & & & & & C_{66} \end{bmatrix} \begin{bmatrix} \varepsilon_1 \\ \varepsilon_2 \\ \varepsilon_3 \\ \varepsilon_4 \\ \varepsilon_5 \\ \varepsilon_6 \end{bmatrix} .
$$
(6.9)

Henceforth only the elements both on the main diagonal and in the upper triangle of the symmetric matrices are given, without the indication of symmetry.

Hooke's law is formulated with respect to an arbitrary coordinate system. Stress and strain components can be transformed to another coordinate system, as can the compliances and stiffness constants. General transformation formulae are given in textbooks such as those listed in section 6.10.

6.3 MATERIAL SYMMETRY

Material can show symmetry with respect to different planes. When at a point P of a material there is a plane of symmetry, A (figure 6.3), that means that a property - for instance the modulus of elasticity - in an arbitrary direction $\{r\}$ through P has the same value as in the direction of the reflection $\{r'\}$ of $\{r\}$ with respect to A.

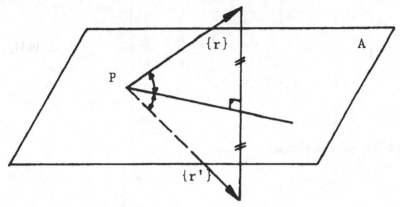

Figure 6.3. Material symmetry at point P with respect to plane A: properties in any direction $\{r\}$ and in the direction of its reflection with respect to A, $\{r'\}$, are equal.

A monoclinic material has at every point one plane of symmetry, and generally all such planes are parallel to each other. Choose a point as origin and the companion plane of symmetry as the x_1x_2-coordinate plane. It can be proved that with respect to the plane of symmetry some of the compliances are zero (see figure 6.4):

$$[S] = \begin{bmatrix} S_{11} & S_{12} & S_{13} & 0 & 0 & S_{16} \\ & S_{22} & S_{23} & 0 & 0 & S_{26} \\ & & S_{33} & 0 & 0 & S_{36} \\ & & & S_{44} & S_{45} & 0 \\ & & & & S_{55} & 0 \\ & & & & & S_{66} \end{bmatrix} . \qquad (6.10)$$

Only 13 independent elastic constants determine the linear elastic behaviour of monoclinic material. After transformation to an arbitrary coordinate system the zeros of the compliance matrix are changed to non-zero values.

In the case of orthotropic material there are at every point three orthogonal planes of symmetry. The directions of the three intersection lines are the principal directions of the material and can be used as the directions of a coordinate system. With respect to this coordinate system the compliance matrix is:

$$[S] = \begin{bmatrix} S_{11} & S_{12} & S_{13} & 0 & 0 & 0 \\ & S_{22} & S_{23} & 0 & 0 & 0 \\ & & S_{33} & 0 & 0 & 0 \\ & & & S_{44} & 0 & 0 \\ & & & & S_{55} & 0 \\ & & & & & S_{66} \end{bmatrix}, \qquad (6.11)$$

and the stiffness matrix:

$$[C] = \begin{bmatrix} C_{11} & C_{12} & C_{13} & 0 & 0 & 0 \\ & C_{22} & C_{23} & 0 & 0 & 0 \\ & & C_{33} & 0 & 0 & 0 \\ & & & C_{44} & 0 & 0 \\ & & & & C_{55} & 0 \\ & & & & & C_{66} \end{bmatrix}. \qquad (6.12)$$

The number of independent elastic constants is reduced to 9.

Working out of the inverse relation between [S] and [C] of orthotropic material results in:

$$C_{11} = \frac{S_{22}S_{33} - S_{23}S_{32}}{\Delta_S} \ ; \quad C_{22} = \frac{S_{33}S_{11} - S_{31}S_{13}}{\Delta_S} \ ;$$

$$C_{33} = \frac{S_{11}S_{22} - S_{12}S_{21}}{\Delta_S} \ ; \quad C_{12} = \frac{S_{13}S_{32} - S_{12}S_{33}}{\Delta_S} \ ;$$

$$C_{23} = \frac{S_{21}S_{13} - S_{23}S_{11}}{\Delta_S} \ ; \quad C_{31} = \frac{S_{32}S_{21} - S_{31}S_{22}}{\Delta_S} \ ; \qquad (6.13)$$

$$C_{44} = \frac{1}{S_{44}} \ ; \quad C_{55} = \frac{1}{S_{55}} \ ; \quad C_{66} = \frac{1}{S_{66}} \ ;$$

where $\Delta_S = S_{11}S_{22}S_{33} - S_{11}S_{23}S_{32} - S_{12}S_{21}S_{33} +$

$$+ S_{12}S_{23}S_{31} + S_{13}S_{21}S_{32} - S_{13}S_{22}S_{31} .$$

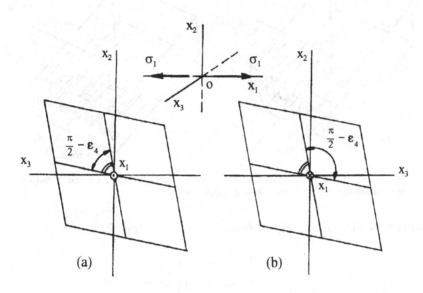

(a)   (b)

Figure 6.4. Uniaxial stress in x_1-direction acting in a material having symmetry with respect to the x_1x_2-plane. The assumption that this uniaxial stress should cause a shear strain $\varepsilon_4 \neq 0$ creates a dilemma: by reason of the symmetry the deformed x_2x_3-plane must look equally observed from both directions, but in one case (a.) $\varepsilon_4 > 0$, and in the other (b.) $\varepsilon_4 < 0$. Conclusion: $\varepsilon_4 = 0$, and thus $S_{41} = 0$.

A layer reinforced by a woven fabric with a plain weave can approximately act as an example of an orthotropic lamina, see figure 6.5b. The two thread directions of the fabric are principal directions of the lamina. A plain weave provides that the half of warp and weft, respectively, lays on one side of the fabric, and promotes globally also a symmetric behaviour in the direction perpendicular to the midplane of the lamina.

Mostly the fibre cross-sections in the transverse plane of a unidirectional FRP are randomly distributed as shown in figure 6.5a. Then in the transverse plane the properties of the material are independent of direction; so that the material is transversely isotropic. All planes parallel to the fibre direction are planes of symmetry, and so is the transverse plane. There remain five independent elastic constants. If the x_1-axis is in the longitudinal direction, and so the x_2x_3-plane corresponds with the transverse plane, the compliance

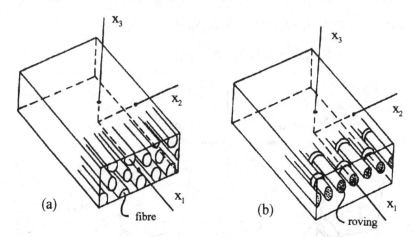

Figure 6.5. A part of a unidirectional layer, microscopically seen (a.), and a part of a layer reinforced with a biased woven fabric (b.).

matrix of the transversely isotropic material has the form

$$[S] = \begin{bmatrix} S_{11} & S_{12} & S_{12} & 0 & 0 & 0 \\ & S_{22} & S_{23} & 0 & 0 & 0 \\ & & S_{22} & 0 & 0 & 0 \\ & & & 2(S_{22}-S_{23}) & 0 & 0 \\ & & & & S_{66} & 0 \\ & & & & & S_{66} \end{bmatrix}, \quad (6.14)$$

and the stiffness matrix is

$$[C] = \begin{bmatrix} C_{11} & C_{12} & C_{12} & 0 & 0 & 0 \\ & C_{22} & C_{23} & 0 & 0 & 0 \\ & & C_{22} & 0 & 0 & 0 \\ & & & \frac{1}{2}(C_{22}-C_{23}) & 0 & 0 \\ & & & & C_{66} & 0 \\ & & & & & C_{66} \end{bmatrix}. \quad (6.15)$$

In the case of uniaxial stress in fibre direction, σ_1, there is no shearing between the x_1-axis and a direction in the transverse plane, because $S_{51} = S_{61} = 0$. Under influence of an off-axis load unidirectional FRP shows shearing between the loading direction and a direction perpendicular to it, see figure 6.6. The relevant compliance can be computed by a transformation formula, and has a non-zero value.

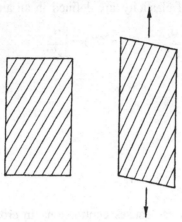

Figure 6.6. Shear strain in a unidirectional lamina due to an off-axis uniaxial stress.

It is well-known that isotropic materials have two independent elastic constants, i.e., two of the three engineering constants - the modulus of elasticity E, the shear modulus G, and Poisson's ratio, v. Only two are independent because there is a relation between these constants:

$$E = 2 \ (1 + v) \ G. \tag{6.16}$$

6.4 ENGINEERING CONSTANTS

Instead of the compliances and the stiffness constants, usually engineering constants are used to characterise the stiffness behaviour of anisotropic materials like FRP. For a clear understanding attention must be paid to the definition of these engineering constants.

The definitions of the moduli of elasticity are based on a situation of one non-zero stress component:

$$\sigma_k \neq 0; \quad \sigma_i = 0 \quad \text{for} \quad i \neq k \quad \text{and} \quad i \in \{1, ..., 6\}. \tag{6.17}$$

The elements of the strain vector caused by σ_k are distinguished by a superscript k in brackets:

$$\varepsilon_i^{(k)} \ ; \quad i \in \{1, ..., 6\}. \tag{6.18}$$

The modulus of elasticity in x_i-direction, E_i, is the quotient of σ_i and $\varepsilon_i^{(i)}$:

$$E_1 = \frac{\sigma_1}{\varepsilon_1^{(1)}} \; ; \quad E_2 = \frac{\sigma_2}{\varepsilon_2^{(2)}} \; ; \quad E_3 = \frac{\sigma_3}{\varepsilon_3^{(3)}} \; . \tag{6.19a}$$

The shear moduli of elasticity are defined in an analogical way:

$$E_4 \; (= G_{23}) = \frac{\sigma_4}{\varepsilon_4^{(4)}} \; ; \quad E_5 \; (= G_{31}) = \frac{\sigma_5}{\varepsilon_5^{(5)}} \; ;$$

$$\tag{6.19b}$$

$$E_6 \; (= G_{12}) = \frac{\sigma_6}{\varepsilon_6^{(6)}} \; .$$

Summarising:

$$E_i = \frac{\sigma_i}{\varepsilon_i^{(i)}} \; ; \quad i \in \{1,...,6\} \; . \tag{6.19}$$

A uniaxial tensile stress causes contractions in cross directions, e.g., σ_1 causes strain $\varepsilon_1^{(1)}$ in x_1-direction and (negative) strains $\varepsilon_1^{(2)}$ and $\varepsilon_1^{(3)}$ in x_2 and x_3-direction, respectively. The Poisson's ratio ν_{12} is the quotient of cross contraction $-\varepsilon_1^{(2)}$ and strain $\varepsilon_1^{(1)}$ in tensile direction:

$$\nu_{12} = -\frac{\varepsilon_2^{(1)}}{\varepsilon_1^{(1)}} \; ; \quad \nu_{13} = -\frac{\varepsilon_3^{(1)}}{\varepsilon_1^{(1)}} \; ; \quad (\text{etc.:} \; 1 \to 2 \to 3 \to 1) \; . \tag{6.20a}$$

The subscripts of ν_{ij} indicate the sequence "cause and effect". Formulae for other Poisson's ratios can be found by cyclic interchange of the subscripts and superscripts 1, 2, and 3 in equation (6.20a).

Anisotropic materials show other effects. A uniaxial stress σ_1 can cause shear strains $\varepsilon_4^{(1)}$, $\varepsilon_5^{(1)}$, and $\varepsilon_6^{(1)}$, etc. These shear strains can be related to the direct strain in the direction of the only non-zero stress component, and so different strain ratios are defined:

$$\rho_{14} = \frac{\varepsilon_4^{(1)}}{\varepsilon_1^{(1)}} \; ; \quad \rho_{15} = \frac{\varepsilon_5^{(1)}}{\varepsilon_1^{(1)}} \; ; \quad \rho_{16} = \frac{\varepsilon_6^{(1)}}{\varepsilon_1^{(1)}} \; . \tag{6.20b}$$

Generally there are 30 different strain ratios or coefficients of mutual influence:

$$\rho_{ij} = \frac{\varepsilon_j^{(i)}}{\varepsilon_i^{(i)}} \; ; \quad i,j \in \{1,...,6\} \; . \tag{6.20}$$

In these definitions the minus sign is lacking, in contradiction to the usual definition of a Poisson's ratio: $v_{12} = -\rho_{12}$, etc.

6.5 RELATIONS BETWEEN ENGINEERING
CONSTANTS AND STIFFNESS/COMPLIANCE

Substitution of the definitions given in equations (6.19) and (6.20) of the engineering constants into expression (6.4a) of Hooke's law results in the following relations between compliances and engineering constants:

$$S_{ij} = \frac{\rho_{ji}}{E_j} \; ; \; i,j \in \{1,...,6\} \; . \tag{6.21}$$

Symmetry of the compliance matrix, expressed in equation (6.7), leads to a formulation of the 15 reciprocal relations in the engineering constants:

$$\frac{\rho_{ij}}{E_i} = \frac{\rho_{ji}}{E_j} \; ; \; i,j \in \{1,...,6\}, \tag{6.22}$$

the so-called Maxwell relations. These 15 relations between 36 engineering constants (6 moduli of elasticity and 30 strain ratios) leave 21 of them mutually independent.

Henceforth the theory is restricted to orthotropic materials. With respect to the principal directions of an orthotropic material the compliance matrix is simplified to:

$$[S] = \begin{bmatrix} \dfrac{1}{E_1} & -\dfrac{v_{21}}{E_2} & -\dfrac{v_{31}}{E_3} & 0 & 0 & 0 \\[2mm] & \dfrac{1}{E_2} & -\dfrac{v_{32}}{E_3} & 0 & 0 & 0 \\[2mm] & & \dfrac{1}{E_3} & 0 & 0 & 0 \\[2mm] & & & \dfrac{1}{G_{23}} & 0 & 0 \\[2mm] & & & & \dfrac{1}{G_{31}} & 0 \\[2mm] & & & & & \dfrac{1}{G_{12}} \end{bmatrix} \; . \tag{6.23}$$

The resulting 9 main engineering constants are defined with respect to the principal directions of the material.

The stiffness constants can be expressed in the engineering constants by substitution of equation (6.23) into equation (6.13):

$$C_{11} = \frac{1 - v_{23} v_{32}}{\Delta} E_1 \; ; \; (\text{etc.}: \; 1 \to 2 \to 3 \to 1) \; ;$$

$$C_{12} = \frac{v_{21} + v_{23} v_{31}}{\Delta} E_1 = C_{21} = \frac{v_{12} + v_{13} v_{32}}{\Delta} E_2 \; ; \; (1 \to 2 \to 3 \to 1) \; ;$$

(6.24)

$$C_{44} = G_{23} \; ; \; C_{55} = G_{31} \; ; \; C_{66} = G_{12} \; ;$$

where $\Delta = 1 - v_{23} v_{32} - v_{31} v_{13} - v_{12} v_{21} - v_{23} v_{31} v_{12} - v_{32} v_{21} v_{13}$.

$(1 \to 2 \to 3 \to 1)$ indicates the valid cyclic interchanges of these subscripts.

Transversely isotropic material is a special case of an orthotropic material. Formulae deduced for orthotropic material can be used for transversely isotropic material. If the $x_2 x_3$-plane coincides with the transverse plane of isotropy, the subscripts 2 and 3 can be replaced with the same subscript: T (transverse direction). Furthermore, subscript 1 can be replaced with L (longitudinal direction).

In connection with unidirectional FRP the following main engineering constants are usual as shown in figure 6.7:

- the longitudinal modulus of elasticity, E_L, in fibre direction
- the transverse modulus of elasticity, E_T, in a direction perpendicular to the fibres
- the longitudinal-transverse (or in-plane) shear modulus, G_{LT}, in a plane parallel to the fibres
- the transverse shear modulus, G_{TT}, in a plane perpendicular to the fibres
- the longitudinal-transverse (or major) Poisson's ratio, v_{LT}
- the transverse-longitudinal (or minor) Poisson's ratio, v_{TL}
- the transverse Poisson's ratio, v_{TT}

There are two relations between the seven main engineering constants for transversely isotropic material. Firstly, the reciprocal relation

$$\frac{v_{LT}}{E_L} = \frac{v_{TL}}{E_T} \; ,$$

(6.25)

and secondly, a relation between the engineering constants which characterise the isotropic behaviour in the transverse plane:

$$E_T = 2(1+\nu_{TT})\,G_{TT}. \qquad (6.26)$$

The latter can be deduced from substitution of equation (6.24) after changing the subscripts into the expression of C_{44} equation (6.15). Compare equation (6.26) with the well-known relation (6.16) for isotropic materials.

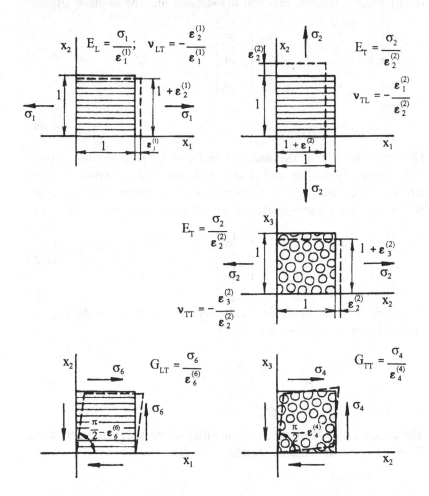

Figure 6.7. Definition of the different engineering constants of transversely isotropic (unidirectional) material of which the longitudinal direction is chosen as x_1-axis.

6.6 PLANE STRESS

Classical laminate theory is based on the assumption that there are plane stress conditions in the different laminae. Let the x_1x_2-plane correspond with the plane of plane stress. That means:

$$\sigma_3 = \sigma_4 = \sigma_5 = 0. \tag{6.27}$$

The in-plane deformation of the x_1x_2-plane is specified by the strain components ε_1, ε_2, and ε_6. Substitution of equation (6.27) in equation (6.4b) gives a relation between the stresses and the in-plane strains:

$$\begin{bmatrix} \varepsilon_1 \\ \varepsilon_2 \\ \varepsilon_6 \end{bmatrix} = \begin{bmatrix} S_{11} & S_{12} & S_{16} \\ & S_{22} & S_{26} \\ & & S_{66} \end{bmatrix} \begin{bmatrix} \sigma_1 \\ \sigma_2 \\ \sigma_6 \end{bmatrix}. \tag{6.28a}$$

In short:

$$\{\underline{\varepsilon}\} = [\underline{S}]\,\{\underline{\sigma}\}. \tag{6.28b}$$

$[\underline{S}]$ is a submatrix of $[S]$, and $\{\underline{\varepsilon}\}$ and $\{\underline{\sigma}\}$ are partitions of $\{\varepsilon\}$ and $\{\sigma\}$, respectively. The underlining indicates the condition of plane stress. Notice that $[\underline{S}]$ is a symmetric 3x3 matrix. Generally the remaining strain components are not zero, but they are less relevant.

$$\begin{bmatrix} \varepsilon_3 \\ \varepsilon_4 \\ \varepsilon_5 \end{bmatrix} = \begin{bmatrix} S_{31} & S_{32} & S_{36} \\ S_{41} & S_{42} & S_{46} \\ S_{51} & S_{52} & S_{56} \end{bmatrix} \begin{bmatrix} \sigma_1 \\ \sigma_2 \\ \sigma_6 \end{bmatrix}. \tag{6.29}$$

The plane stress components can be expressed in terms of the in-plane strains by inverting equation (6.28):

$$\begin{bmatrix} \sigma_1 \\ \sigma_2 \\ \sigma_6 \end{bmatrix} = \begin{bmatrix} Q_{11} & Q_{12} & Q_{16} \\ & Q_{22} & Q_{26} \\ & & Q_{66} \end{bmatrix} \begin{bmatrix} \varepsilon_1 \\ \varepsilon_2 \\ \varepsilon_6 \end{bmatrix}. \tag{6.30}$$

The coefficients Q_{ij} are the reduced stiffness constants for plane stress. They have the dimension of stress. It holds:

$$[\underline{Q}] = [\underline{S}]^{-1}. \tag{6.31}$$

The reduced stiffness matrix $[Q]$ is a symmetric one, because $[\underline{S}]$ is symmetric.

$$Q_{ji} = Q_{ij} \; ; \quad i,j \in \{1,2,6\}. \tag{6.32}$$

In the case of orthotropy and with respect to the principal directions of the material, equation (6.30) can be simplified:

$$\begin{bmatrix} \sigma_1 \\ \sigma_2 \\ \sigma_6 \end{bmatrix} = \begin{bmatrix} Q_{11} & Q_{12} & 0 \\ & Q_{22} & 0 \\ & & Q_{66} \end{bmatrix} \begin{bmatrix} \varepsilon_1 \\ \varepsilon_2 \\ \varepsilon_6 \end{bmatrix} . \qquad (6.33)$$

Only four independent elastic constants play an important part. The reduced stiffnesses are related to the main engineering constants:

$$Q_{11} = \frac{E_1}{1 - \nu_{12}\nu_{21}} \;\; ; \;\; Q_{22} = \frac{E_2}{1 - \nu_{12}\nu_{21}} \; ;$$

$$Q_{12} = \frac{\nu_{21}E_1}{1 - \nu_{12}\nu_{21}} = Q_{21} = \frac{\nu_{12}E_2}{1 - \nu_{12}\nu_{21}} \; ; \qquad (6.34)$$

$$Q_{66} = G_{12} \; ,$$

as a result of the substitution of equation (6.23) into equation (6.31).

6.7 RESTRICTIONS ON ELASTIC CONSTANTS

To bring an elastic material from a state without stress into an arbitrary stressed state, positive energy input is required to obey the laws of physics. The amount of energy required should be independent of the order in which loads are applied to reach this stressed state. The energy input per unit volume of the material is called the specific strain energy, U, corresponding to the particular state of stress.

In the case of linear elasticity the specific strain energy is a function of the stress and strain components of this particular state of stress, given by:

$$U = \frac{1}{2} \sum_{i=1}^{6} \sigma_i \, \varepsilon_i = \frac{1}{2} \{\sigma^T\}\{\varepsilon\} = \frac{1}{2} \{\varepsilon^T\}\{\sigma\} > 0 . \qquad (6.35)$$

Substitution of Hooke's law, equations (6.4c) and (6.5), into equation (6.35) yields:

$$U = \frac{1}{2}\{\sigma^T\}[S]\{\sigma\} = \frac{1}{2}\{\varepsilon^T\}[C]\{\varepsilon\} > 0. \qquad (6.36)$$

In the case of one non-zero stress component, σ_i, the inequality equation (6.36) is reduced to:

$$S_{ii}\sigma_i^2 > 0.$$

It will be clear that every diagonal element of [S] must be positive. And so also diagonal elements of [C].

$$S_{ii} > 0 \; ; \quad C_{ii} > 0 \; ; \quad i \in \{1,...,6\}.$$ (6.37a)

Using equation (6.21) the restrictions can be expressed in terms of the moduli of elasticity:

$$E_i > 0 \; ; \quad E_1, E_2, E_3, G_{23}, G_{31}, G_{12} > 0.$$ (6.37b)

In the case of only two non-zero stress components, σ_i and σ_j, inequality equation (6.36) leads to:

$$S_{ii} \sigma_i^2 + 2S_{ij} \sigma_i \sigma_j + S_{jj} \sigma_j^2 > 0.$$

After rearranging:

$$S_{ii} \{\sigma_i + \frac{S_{ij}}{S_{ii}} \sigma_j\}^2 + \{S_{jj} - \frac{S_{ij}^2}{S_{ii}} \} \sigma_j^2 > 0.$$

In combination with equation (6.37) it leads to the restriction

$$S_{ii} S_{jj} - S_{ij}^2 > 0.$$ (6.38)

Substitution of equation (6.21) into equation (6.38) gives restrictions on the strain ratios in pairs:

$$\rho_{ij} \, \rho_{ji} < 1,$$ (6.39)

and further, with the help of equation (6.22):

$$\rho_{ij}^2 < \frac{E_i}{E_j} \; ; \quad i,j \in \{1,...,6\}.$$ (6.40)

These restrictions applied to the Poisson's ratios give:

$$\nu_{ij} \, \nu_{ji} < 1; \quad i,j \in \{1,2,3\}.$$ (6.41)

$$\nu_{ij}^2 < \frac{E_i}{E_j}.$$ (6.42)

The exercise can be continued for all combinations of three non-zero stress components, etc.

In the case of plane stress there are the same restrictions as equations (6.37) and (6.38) on the reduced stiffnesses:

$$Q_{ii} > 0 \; ; \quad Q_{ii}Q_{jj} - Q_{ij}^2 > 0 \; ; \quad i,j \in \{1,2,6\}.$$ (6.43)

So far no use has been made of material symmetry. Lempriere derived narrower restrictions on the Poisson's ratios of orthotropic materials. With the help of equations (6.37) and (6.41), the expression of C_{11} in equation (6.24) leads to the restriction

$$\Delta = 1 - v_{23}v_{32} - v_{31}v_{13} - v_{12}v_{21} - v_{23}v_{31}v_{12} - v_{32}v_{21}v_{13} > 0. \qquad (6.44)$$

And using the reciprocal relations (6.22), expressed in Poisson's ratios, it leads to:

$$2\,v_{23}v_{31}v_{12} < 1 - \frac{E_3}{E_2}v_{23}^2 - \frac{E_1}{E_3}v_{31}^2 - \frac{E_2}{E_1}v_{12}^2 < 1.$$

The latter inequality is a consequence of the moduli of elasticity being positive. Restrictions on v_{12} can be formulated as an appropriate solution:

$$-\left\{ \frac{E_1}{E_2}(1 - \frac{E_3}{E_2}v_{23}^2)(1 - \frac{E_1}{E_3}v_{31}^2) \right\}^{\frac{1}{2}} - \frac{E_1}{E_2}v_{23}v_{31} < v_{12} <$$

$$\qquad\qquad (6.45)$$

$$\left\{ \frac{E_1}{E_2}(1 - \frac{E_3}{E_2}v_{23}^2)(1 - \frac{E_1}{E_3}v_{31}^2) \right\}^{\frac{1}{2}} - \frac{E_1}{E_2}v_{23}v_{31}; (etc.: 1\to2\to3\to1).$$

Cyclic interchange of the subscripts leads to restrictions on v_{23} and v_{31}.

Application of equation (6.42) to transversely isotropic materials gives the restrictions

$$v_{LT}^2 < \frac{E_L}{E_T} \;;\; v_{TL}^2 < \frac{E_T}{E_L} \;;\; v_{TT}^2 < 1 \;;$$

and of equation (6.44):

$$\Delta = (1 + v_{TT})(1 - v_{TT} - 2v_{LT}v_{TL}) > 0.$$

Then it holds:

$$-1 < v_{TT} < 1 - 2v_{LT}v_{TL} = 1 - 2\frac{E_T}{E_L}v_{LT}^2. \qquad (6.46)$$

$$v_{LT}^2 < \frac{1}{2}\frac{E_L}{E_T}(1 - v_{TT}) \;;\; v_{TL}^2 < \frac{1}{2}\frac{E_T}{E_L}(1 - v_{TT}). \qquad (6.47)$$

Strongly anisotropic materials with $E_L \gg E_T$ have a major Poisson's ratio, v_{LT}, larger than 1.

6.8 TRANSFORMATION OF ELASTIC CONSTANTS IN A PLANE

If in a point of a material the elastic constants are known with respect to a given coordinate system with axes x_1, x_2, x_3, the values of the elastic constants with respect to another coordinate system, x_1', x_2', x_3' can be derived by transformation.

In this context transformations shall be limited to rotations in the plane of a lamina, in preparation to the laminate theory. The $x_1 x_2$-plane and the $x_1' x_2'$-plane coincide with the midplane of the lamina. Thus, the x_3'-axis coincides with the x_3-axis. The positive transformation angle is φ, see figure 6.8.

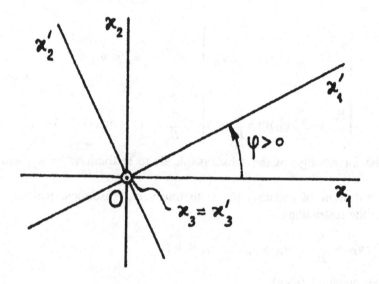

Figure 6.8. The positive direction of transformation angle φ at rotation around the x_3-axis.

The transformation equation for the stress components in the case of plane stress in the $x_1 x_2$-plane is shown in figure 6.9:

$$\begin{bmatrix} \sigma_1' \\ \sigma_2' \\ \sigma_6' \end{bmatrix} = \begin{bmatrix} \cos^2\varphi & \sin^2\varphi & 2\sin\varphi\cos\varphi \\ \sin^2\varphi & \cos^2\varphi & -2\sin\varphi\cos\varphi \\ -\sin\varphi\cos\varphi & \sin\varphi\cos\varphi & \cos^2\varphi - \sin^2\varphi \end{bmatrix} \begin{bmatrix} \sigma_1 \\ \sigma_2 \\ \sigma_6 \end{bmatrix}. \tag{6.48a}$$

In short:

$$\{\underline{\sigma}'\} = [T]\,\{\underline{\sigma}\}\,. \tag{6.48b}$$

[T] is the transformation matrix.

In established laminate practice the negative transformation angle θ (= $-\varphi$) is used to indicate the orientation of the lamina with respect to the main axes of the laminate. For that reason the required transformation equations are given as functions of θ instead of φ. Call

$$m = \cos\theta \;\; ; \;\; n = \sin\theta. \tag{6.49}$$

Transformation equation (6.48) can now be expressed in orientation angle θ:

$$\begin{bmatrix} \sigma_1' \\ \sigma_2' \\ \sigma_6' \end{bmatrix} = \begin{bmatrix} m^2 & n^2 & -2mn \\ n^2 & m^2 & 2mn \\ mn & -mn & m^2-n^2 \end{bmatrix} \begin{bmatrix} \sigma_1 \\ \sigma_2 \\ \sigma_6 \end{bmatrix}. \tag{6.48c}$$

In equation (6.48c) [T] is a function of θ.

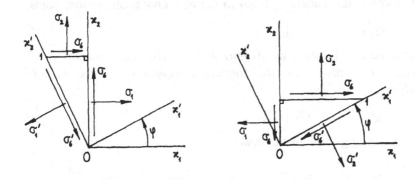

Figure 6.9. Plane stress in the x_1x_2-plane. A new coordinate system $x_1'x_2'x_3'$ after rotation around the x_3-axis. The stress transformation equations can be derived from equilibrium of the sketched small blocks.

If instead of the engineering shear strains defined by equation (6.3) the tensor shear strains (being the half of the engineering shear strains, respectively) are used, the transformation equations for the stresses hold also for the strains:

$$\begin{bmatrix} \varepsilon_1' \\ \varepsilon_2' \\ \frac{1}{2}\varepsilon_6' \end{bmatrix} = \begin{bmatrix} m^2 & n^2 & -2mn \\ n^2 & m^2 & 2mn \\ mn & -mn & m^2-n^2 \end{bmatrix} \begin{bmatrix} \varepsilon_1 \\ \varepsilon_2 \\ \frac{1}{2}\varepsilon_6 \end{bmatrix} . \tag{6.50}$$

And so:

$$\begin{bmatrix} \varepsilon_1' \\ \varepsilon_2' \\ \varepsilon_6' \end{bmatrix} = \begin{bmatrix} m^2 & n^2 & -mn \\ n^2 & m^2 & mn \\ 2mn & -2mn & m^2-n^2 \end{bmatrix} \begin{bmatrix} \varepsilon_1 \\ \varepsilon_2 \\ \varepsilon_6 \end{bmatrix} . \tag{6.51a}$$

In short:

$$\{\underline{\varepsilon}'\} = [T^T]^{-1} \{\underline{\varepsilon}\} , \tag{6.51b}$$

because the transformation matrix in equation (6.51a) is the inverse of the transposed [T]. In the case of plane stress, the transformation of Hooke's law equation (6.30) to the new coordinate system reads:

$$\{\underline{\sigma}'\} = [Q'] \{\underline{\varepsilon}'\} . \tag{6.52}$$

Substitution of equations (6.48b) and (6.51b) into equation (6.52) gives as a result the transformation equation for the reduced stiffnesses:

$$[Q'] = [T] [Q] [T^T] . \tag{6.53a}$$

After working through the details:

$$\begin{bmatrix} Q_{11}' \\ Q_{22}' \\ Q_{12}' \\ Q_{66}' \\ Q_{16}' \\ Q_{26}' \end{bmatrix} = \begin{bmatrix} m^4 & n^4 & 2m^2n^2 & 4m^2n^2 & -4m^3n & -4mn^3 \\ n^4 & m^4 & 2m^2n^2 & 4m^2n^2 & 4mn^3 & 4m^3n \\ m^2n^2 & m^2n^2 & m^4+n^4 & -4m^2n^2 & 2mn(m^2-n^2) & -2mn(m^2-n^2) \\ m^2n^2 & m^2n^2 & -2m^2n^2 & (m^2-n^2)^2 & 2mn(m^2-n^2) & -2mn(m^2-n^2) \\ m^3n & -mn^3 & -mn(m^2-n^2) & -2mn(m^2-n^2) & m^2(m^2-3n^2) & n^2(3m^2-n^2) \\ mn^3 & -m^3n & mn(m^2-n^2) & 2mn(m^2-n^2) & n^2(3m^2-n^2) & m^2(m^2-3n^2) \end{bmatrix} \begin{bmatrix} Q_{11} \\ Q_{22} \\ Q_{12} \\ Q_{66} \\ Q_{16} \\ Q_{26} \end{bmatrix} . \tag{6.53b}$$

With respect to the axes of symmetry of an orthotropic lamina:

$$Q_{16} = Q_{26} = 0 .$$

Substitution into equation (6.53b) reduces the transformation equations to

$$
\begin{bmatrix} Q'_{11} \\ Q'_{22} \\ Q'_{12} \\ Q'_{66} \\ Q'_{16} \\ Q'_{26} \end{bmatrix} = \begin{bmatrix} m^4 & n^4 & 2m^2n^2 & 4m^2n^2 \\ n^4 & m^4 & 2m^2n^2 & 4m^2n^2 \\ m^2n^2 & m^2n^2 & m^4+n^4 & -4m^2n^2 \\ m^2n^2 & m^2n^2 & -2m^2n^2 & (m^2-n^2)^2 \\ m^3n & -mn^3 & -mn(m^2-n^2) & -2mn(m^2-n^2) \\ mn^3 & -m^3n & mn(m^2-n^2) & 2mn(m^2-n^2) \end{bmatrix} \cdot \begin{bmatrix} Q_{11} \\ Q_{22} \\ Q_{12} \\ Q_{66} \end{bmatrix} \qquad (6.54)
$$

The following trigonometric relations:

$$\cos^4\theta = \tfrac{1}{8}\,(3 + 4\cos 2\theta + \cos 4\theta)\ ;$$

$$\cos^3\theta\,\sin\theta = \tfrac{1}{8}\,(2\sin 2\theta + \sin 4\theta)\ ;$$

$$\cos^2\theta\,\sin^2\theta = \tfrac{1}{8}\,(1 - \cos 4\theta)\ ; \qquad\qquad (6.55)$$

$$\cos\theta\,\sin^3\theta = \tfrac{1}{8}\,(2\sin 2\theta - \sin 4\theta)\ ;$$

$$\sin^4\theta = \tfrac{1}{8}\,(3 - 4\cos 2\theta + \cos 4\theta)\ ,$$

can be used to rewrite the transformation equations (6.54) in terms of linear trigonometric functions:

$$Q'_{11} = U_1 + U_2\cos 2\theta + U_3\cos 4\theta\ ;$$

$$Q'_{22} = U_1 - U_2\cos 2\theta + U_3\cos 4\theta\ ;$$

$$Q'_{12} = U_4 - U_3\cos 4\theta\ ;$$

$$\qquad\qquad (6.56)$$

$$Q'_{66} = U_5 - U_3\cos 4\theta\ ;$$

$$Q'_{16} = \tfrac{1}{2}\,U_2\sin 2\theta + U_3\sin 4\theta\ ;$$

$$Q'_{26} = \tfrac{1}{2}\,U_2\sin 2\theta - U_3\sin 4\theta\ ,$$

where

$$U_1 = \frac{1}{8} (3Q_{11} + 3Q_{22} + 2Q_{12} + 4Q_{66}) \ ;$$
$$U_2 = \frac{1}{2} (Q_{11} - Q_{22}) \ ;$$
$$U_3 = \frac{1}{8} (Q_{11} + Q_{22} - 2Q_{12} - 4Q_{66}) \ ; \qquad (6.57)$$
$$U_4 = \frac{1}{8} (Q_{11} + Q_{22} + 6Q_{12} - 4Q_{66}) \ ;$$
$$U_5 = \frac{1}{8} (Q_{11} + Q_{22} - 2Q_{12} + 4Q_{66}) \ .$$

It turns out that the following quantities - linear combinations of Q_{ij} -

$$L_1 = Q'_{11} + Q'_{22} + 2Q'_{12} = 2(U_1 + U_4) = Q_{11} + Q_{22} + 2Q_{12} \ ;$$
$$\qquad\qquad (6.58)$$
$$L_2 = Q'_{66} - Q'_{12} = U_5 - U_4 = Q_{66} - Q_{12} \ ,$$

and also

$$U_1 = \frac{1}{8} (3L_1 + 4L_2) \ ;$$
$$U_4 = \frac{1}{8} (L_1 - 4L_2) \ ; \qquad (6.59)$$
$$U_5 = \frac{1}{2} (U_1 - U_4) = \frac{1}{8} (L_1 + 4L_2) \ ,$$

are invariant in relation to rotation in the plane of the lamina.

The dependency of the reduced stiffnesses of an orthotropic lamina on the transformation angle φ ($= -\theta$) is illustrated in figure 6.10. The chosen example is a unidirectional glass/polyester of a fibre-volume fraction $v_f \approx 0.6$.

To transform the compliance matrix $[\underline{S}]$ substitute equation (6.31) into equation (6.53). The result is:

$$[\underline{S'}] = [T^T]^{-1} \ [\underline{S}] \ [T]^{-1} \ . \qquad (6.60)$$

Substitution of equation (6.21) into equation (6.60) gives the transformation equations for the engineering constants:

$$
\begin{bmatrix} 1/E'_1 \\ 1/E'_2 \\ \rho'_{12}/E'_1 \\ 1/E'_6 \\ \rho'_{16}/E'_1 \\ \rho'_{26}/E'_2 \end{bmatrix}
=
\begin{bmatrix}
m^4 & n^4 & 2m^2n^2 & m^2n^2 & -2m^3n & -2mn^3 \\
n^4 & m^4 & 2m^2n^2 & m^2n^2 & 2mn^3 & 2m^3n \\
m^2n^2 & m^2n^2 & m^4+n^4 & -m^2n^2 & mn(m^2-n^2) & -mn(m^2-n^2) \\
4m^2n^2 & 4m^2n^2 & -8m^2n^2 & (m^2-n^2)^2 & 4mn(m^2-n^2) & -4mn(m^2-n^2) \\
2m^3n & -2mn^3 & -2mn(m^2-n^2) & -mn(m^2-3n^2) & m^2(m^2-3n^2) & n^2(3m^2-n^2) \\
2mn^3 & -2m^3n & 2mn(m^2-n^2) & mn(m^2-n^2) & n^2(3m^2-n^2) & m^2(m^2-3n^2)
\end{bmatrix}
\begin{bmatrix} 1/E_1 \\ 1/E_2 \\ \rho_{12}/E_1 \\ 1/E_6 \\ \rho_{16}/E_1 \\ \rho_{26}/E_2 \end{bmatrix}
$$

$$\qquad\qquad (6.61)$$

In the case of an orthotropic lamina and with respect to the axes of symmetry it holds that:

$$\rho_{16} = \rho_{26} = 0 \ .$$

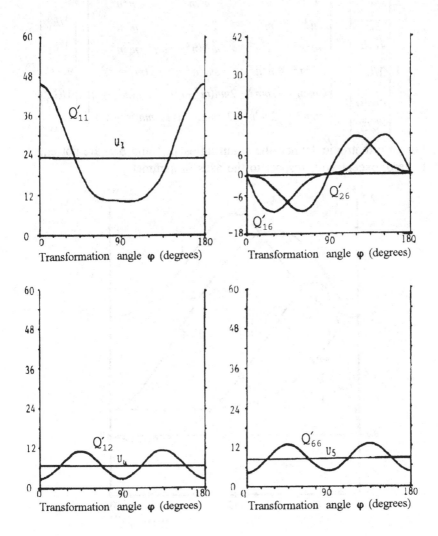

Figure 6.10. The reduced stiffness for plane stress, Q_{ij}, of an orthotropic lamina (unidirectional glass/polyester with fibre-volume fraction $v_f \approx 0.6$) showing variation with the transformation angle φ. (Units for ordinates are GN/m^2.)

Then equation (6.61) reduces to:

$$
\begin{bmatrix}
1/E_1' \\[4pt]
1/E_2' \\[4pt]
v_{12}'/E_1' \\[4pt]
1/G_{12}' \\[4pt]
\rho_{16}'/E_1' \\[4pt]
\rho_{26}'/E_2'
\end{bmatrix}
=
\begin{bmatrix}
m^4 & n^4 & -2m^2n^2 & m^2n^2 \\[4pt]
n^4 & m^4 & -2m^2n^2 & m^2n^2 \\[4pt]
-m^2n^2 & -m^2n^2 & m^4+n^4 & m^2n^2 \\[4pt]
4m^2n^2 & 4m^2n^2 & 8m^2n^2 & (m^2-n^2)^2 \\[4pt]
2m^3n & -2mn^3 & 2mn(m^2-n^2) & -mn(m^2-n^2) \\[4pt]
2mn^3 & -2m^3n & -2mn(m^2-n^2) & mn(m^2-n^2)
\end{bmatrix}
\begin{bmatrix}
1/E_1 \\[4pt]
1/E_2 \\[4pt]
v_{12}/E_1 \\[4pt]
1/G_{12}
\end{bmatrix}.
$$

$$(6.62)$$

For an orthotropic lamina the strain ratios ρ_{16}' and ρ_{26}' are not equal to zero, except with respect to the axes of symmetry.

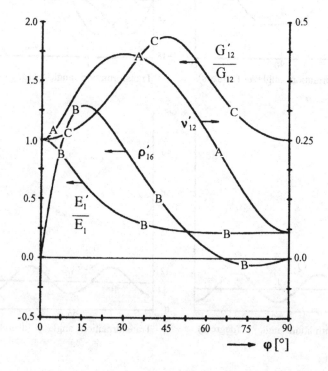

Figure 6.11. The engineering constants of the orthotropic lamina (unidirectional glass/polyester with fibre-volume fraction $v_f \approx 0.6$) showing variation with the transformation angle φ. The moduli E_1' and G_{12}' are plotted on a relative scale, and the ratios ρ_{16}' on an absolute scale.

Figure 6.11 illustrates the dependency of the (relative) engineering constants of an orthotropic lamina on the transformation angle $\varphi (=-\theta)$ using the same example as in figure 6.10.

6.9 STRESS-STRAIN VARIATION IN A LAMINA

A lamina can be considered as a homogeneous anisotropic plate, with a finite thickness. If the loading is limited to forces in the plane, and bending and twisting moments (excluding forces perpendicular to the plane) classical plate theory can be used.

Introduce an orthogonal coordinate system with axes x, y, and z, with the xy-plane coinciding with the midplane of the plate (lamina). The displacements of a point are u, v, and w in the direction of the x-, y- and z-axes, respectively. The displacements of a point in the midplane are distinguished by a superscript 0: u^0, v^0, and w^0.

Suppose that a material line perpendicular to the midplane will, in the deformed state of the plate, stay straight and still perpendicular to the deformed and bent midplane (the Kirchhoff-Love hypothesis). Shear strains between the z-axis and the midplane are neglected, and so are the corresponding shear stresses. Further, the strain in the z-direction is neglected.

Consider pure bending of the x-axis in the xz-plane, in a direction perpendicular to the midplane, see figure 6.12. If the displacements of the x-axis in z-direction are $w^0(x)$, and if the angle of inclination, β, is small it follows that:

$$\beta \approx \tan \beta \; = \; \frac{\partial w^0}{\partial x} \; .$$

Point A is a point on the x-axis: A (x,0,0). Point B has the same x- and y-coordinates as A, and is at a distance z from A. Point A is the projection of point B (x,0,z) onto the midplane.

The displacement in the x-direction of B relative to that of A is:

$$u'' \; = \; -z \sin \beta \; \approx \; -z \tan \beta \; = \; -z \, \frac{\partial w^0}{\partial x} \; .$$

In figure 6.12, inclination angle β is negative. For that reason the minus sign in the last equation is necessary.

The x-displacement of A is u^0. The x-displacement of B is, totally:

$$u \; = \; u^0 + u'' \; = \; u^0 - z \, \frac{\partial w^0}{\partial x} \; . \tag{6.63}$$

The curvature, κ_x, of the deflection curve $w^0(x)$ is the reciprocal of the radius of curvature, R_x. For small deflections:

$$\kappa_x = \frac{1}{R_x} = -\frac{\dfrac{\partial^2 w^0}{\partial x^2}}{\{1 + (\dfrac{\partial w^0}{\partial x})^2\}^{3/2}} \approx -\frac{\partial^2 w^0}{\partial x^2} \,. \qquad (6.64)$$

In figure 6.12 the bending curvature, κ_x, is positive.

Substitution of equations (6.63) and (6.64) into equation (6.3) gives:

$$\varepsilon_x = \frac{\partial u}{\partial x} = \frac{\partial u^0}{\partial x} - z \frac{\partial^2 w^0}{\partial x^2} = \varepsilon_x^0 + z\,\kappa_x \,. \qquad (6.65a)$$

The strain ε_x in B is expressed as a function of the strain ε_x^0 at the projection point A, the bending curvature κ_x of the deflection curve w^0 at A, and the "vertical" distance z from B to A.

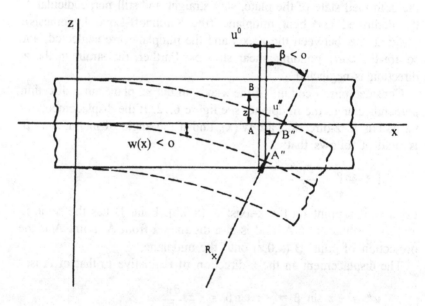

Figure 6.12. Cross-section of a part of a plate along the xz-plane, curved under the influence of loads.

In an analogous way strain ε_y and shear strain ε_{xy} at B can be expressed in terms of the deformations of the midplane in the projection point and the distance to the midplane.

$$\varepsilon_y = \frac{\partial v}{\partial y} = \frac{\partial v^0}{\partial y} - z \frac{\partial^2 w^0}{\partial y^2} = \varepsilon_y^0 + z\,\kappa_y \,. \qquad (6.65b)$$

$$\varepsilon_{xy} = \frac{\partial}{\partial y}\left(u^0 - z\,\frac{\partial w^0}{\partial x}\right) + \frac{\partial}{\partial x}\left(v^0 - z\,\frac{\partial w^0}{\partial y}\right)$$

(6.65c)

$$= \frac{\partial u^0}{\partial y} + \frac{\partial v^0}{\partial x} - 2z\,\frac{\partial^2 w^0}{\partial x\,\partial y} = \varepsilon_{xy}^0 + z\kappa_{xy}\,,$$

where $\varepsilon_{xy}^0 = \dfrac{\partial u^0}{\partial y} + \dfrac{\partial v^0}{\partial x}$; $\kappa_{xy} = -2\,\dfrac{\partial^2 w^0}{\partial x\,\partial y}$.

The curvature κ_{xy} is called the twist curvature.

The strains ε_x, ε_y and ε_{xy} in planes parallel to the midplane can be understood as an in-plane strain vector $\{\underline{\varepsilon}\}$. Hereafter underlining is omitted. The curvatures of the midplane make up a curvature vector $\{\kappa\}$. Use of superscript 0 is not necessary, because curvatures of other planes are not referred to.

$$\{\varepsilon\} = \begin{bmatrix} \varepsilon_x \\ \varepsilon_y \\ \varepsilon_{xy} \end{bmatrix} = \begin{bmatrix} \dfrac{\partial u}{\partial x} \\[2mm] \dfrac{\partial v}{\partial y} \\[2mm] \dfrac{\partial u}{\partial y} + \dfrac{\partial v}{\partial x} \end{bmatrix} \; ; \quad \{\kappa\} = \begin{bmatrix} \kappa_x \\ \kappa_y \\ \kappa_{xy} \end{bmatrix} = \begin{bmatrix} -\dfrac{\partial^2 w^0}{\partial x^2} \\[2mm] -\dfrac{\partial^2 w^0}{\partial y^2} \\[2mm] -2\dfrac{\partial^2 w^0}{\partial x\,\partial y} \end{bmatrix} .$$

(6.66)

Summarising:

$$\{\varepsilon\} = \{\varepsilon^0\} + z\{\kappa\} .$$

(6.65)

To define the loadings of the plate a rectangular part of the plate is cut out parallel to the coordinate planes as shown in figure 6.13. The stresses are integrated over the thickness, h, of the plate. The resultants are forces per unit length of the edge.

The normal stress resultants are N_x and N_y, and the shear stress resultant is N_{xy}.

$$N_x = \int_{-\frac{1}{2}h}^{\frac{1}{2}h} \sigma_x \, dz \; ; \; N_y = \int_{-\frac{1}{2}h}^{\frac{1}{2}h} \sigma_y \, dz \; ; \; N_{xy} = \int_{-\frac{1}{2}h}^{\frac{1}{2}h} \sigma_{xy} \, dz \; . \quad (6.67a)$$

The variation in normal stress component σ_x in the z-direction corresponds to a bending moment per unit length of the edge, M_x, see figure 6.14. The stress components σ_y and σ_{xy} also provide moments per unit length: bending moment M_y and twisting moment M_{xy}, respectively.

$$M_x = \int_{-\frac{1}{2}h}^{\frac{1}{2}h} \sigma_x \, z \, dz \, ; \quad M_y = \int_{-\frac{1}{2}h}^{\frac{1}{2}h} \sigma_y \, z \, dz \, ;$$

$$M_{xy} = \int_{-\frac{1}{2}h}^{\frac{1}{2}h} \sigma_{xy} \, z \, dz \, .$$

(6.68a)

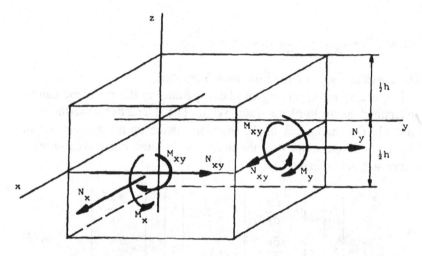

Figure 6.13. Forces and moments per unit length of the edge acting in the plate.

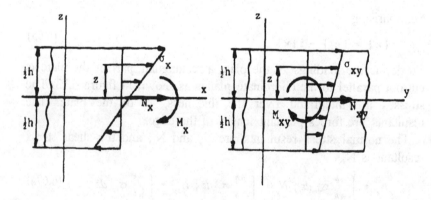

Figure 6.14. Side and frontal view of an edge of the plate parallel to the yz-plane. Forces and moments per unit length are with respect to the midplane.

The introduction of these moments makes it possible to consider that the forces per unit length act at the midplane.

The resultants can be combined to form vectors:

$$\{N\} = \begin{bmatrix} N_x \\ N_y \\ N_{xy} \end{bmatrix} = \int_{-\frac{1}{2}h}^{\frac{1}{2}h} \begin{bmatrix} \sigma_x \\ \sigma_y \\ \sigma_{xy} \end{bmatrix} dz = \int_{-\frac{1}{2}h}^{\frac{1}{2}h} \{\sigma\}\ dz\ . \tag{6.67b}$$

$$\{M\} = \begin{bmatrix} M_x \\ M_y \\ M_{xy} \end{bmatrix} = \int_{-\frac{1}{2}h}^{\frac{1}{2}h} \begin{bmatrix} \sigma_x \\ \sigma_y \\ \sigma_{xy} \end{bmatrix} z\ dz = \int_{-\frac{1}{2}h}^{\frac{1}{2}h} \{\sigma\}\ z\ dz\ . \tag{6.68b}$$

Some textbooks use subscript s for shear, instead of the subscript combination xy and/or 6.

This theory can be used for an orthotropic plate (lamina). The strain distribution as a function of z is given in equation (6.65). Call the reduced stiffness matrix for plane stress with respect to the x, y, z-coordinate system [Q]. Substitution of [Q] and equation (6.65) into equation (6.30) gives the stress distribution as a function of z:

$$\{\sigma\} = \begin{bmatrix} \sigma_x \\ \sigma_y \\ \sigma_{xy} \end{bmatrix} = \begin{bmatrix} Q_{11} & Q_{12} & Q_{16} \\ & Q_{22} & Q_{26} \\ & & Q_{66} \end{bmatrix} \begin{bmatrix} \varepsilon_x \\ \varepsilon_y \\ \varepsilon_{xy} \end{bmatrix} = [Q]\ \{\varepsilon^0\} + z\ [Q]\ \{\kappa\}. \tag{6.69}$$

To know the loads substitute equation (6.69) into equations (6.67b) and (6.68b).

6.10 BIBLIOGRAPHY

1] Lekhnitskii, S.G., "Theory of Elasticity of an Anisotropic Elastic Body", Holden Day, San Francisco, 1963.
2] Sendeckyj, G.P. (ed.), "Mechanics of Composite Materials", Academic Press, New York, 1974.
3] Jones, R.M., "Mechanics of Composite Materials", McGraw-Hill Kogakusha, Tokyo, 1975.
4] Christensen, R.M., "Mechanics of Composite Materials", Wiley, New York, 1979.
5] Agarwal, B.V., Broutman, L.J., "Analysis and Performance of Fiber Composites", Wiley, New York, 1980.
6] Tsai, S.W., Hahn, H.T., "Introduction to Composite Materials", Technomic, Westport, 1980.

7 ANALYSIS OF LAMINATED COMPOSITES

7.1 INTRODUCTION: LAMINATE CONSTRUCTIONS

A laminate is a layered construction shaped as a plate or shell. The layers - laminae - are reinforced with a plane fibre construction like a woven or a non-woven fabric. If the non-woven fabric has a random distribution of the fibres the lamina has a plane isotropic character. In the case of a woven reinforcement the lamina has more or less a (plane) orthotropic character. To predict the stiffness and strength properties of a laminate, it is necessary to know the construction of the laminate and the stiffness and strength properties of the constituent laminae. The construction of the laminate includes the number and kind of the different laminae, and their position and orientation within the laminate.

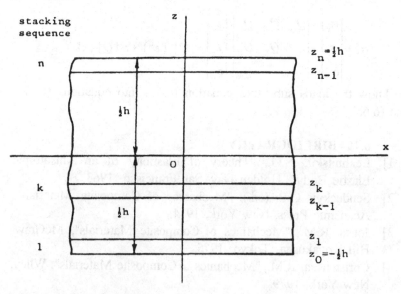

Figure 7.1. The construction of a laminate: stacking sequence of the laminae (ranking bottom-top) and the z-coordinates of the interfaces.

Introduce an orthogonal coordinate system x, y, z. The xy-plane is the midplane of the laminate. If there are axes of symmetry or for other reasons principal directions in the midplane, x- and y-axis are taken in these directions. Call the (constant) thickness of the laminate h. Then the faces of the laminate are $z = -\frac{1}{2}h$ and $z = \frac{1}{2}h$. The number of laminae is n. To indicate the laminae they are numbered from 1 to n in the stacking sequence bottom to top. The interfaces of the lamina with rank k have as z-coordinate z_{k-1} and z_k, respectively, with $z_{k-1} < z_k$, see figure 7.1. So the faces of the laminate are $z_0 = -\frac{1}{2}h$ and $z_n = \frac{1}{2}h$.

The orientation of the kth lamina is described by the orientation angle θ_k between the most important principal direction of the lamina - the $(x_1)_k$-axis - and the x-axis of the laminate. The orientation angle θ_k corresponds to the negative transformation angle, see figure 7.2.

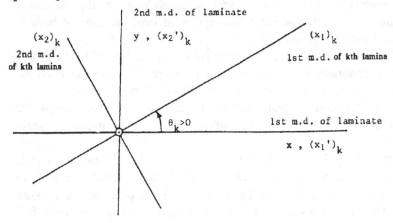

Figure 7.2. The principal (or main) directions (m.d.) of the kth lamina with respect to the laminate m.d. The positive direction of orientation angle θ_k of this lamina.

It is common practice to use a laminate code to indicate the construction of a laminate of identical (orthotropic) laminae. The laminae are denoted by their orientation angle in the sequence bottom to top.

For instance:

[0/0/0/45/45/- 45/- 45/90/90/- 45/- 45/45/45/0/0/0].

Successive laminae with the same orientation can be taken together, the number given as subscript. The above example becomes:

$[0_3/45_2/-45_2/90_2/-45_2/45_2/0_3]$.

In the case of a symmetric construction with respect to the midplane it is sufficient to give the bottom half extended by subscript s. The example is a symmetric laminate, that can be specified by

$$[0_3/45_2/- 45_2/90]_s.$$

For a laminate of different laminae types it is necessary to extend every orientation angle by an indication of the kind of lamina.

7.2 CLASSICAL LAMINATE THEORY

The two basic assumptions of the classical laminated plate theory are: (a) there is a plane stress condition in each of the laminae, (b) there is no slippage between laminae, and a material line perpendicular to the midplane stays straight and perpendicular to the deformed midplane. To fulfil the latter assumption at least the laminae have to be well glued together.

The laminate then deforms as an entity, equation (6.65). The strains change continuously throughout the laminate. On the other hand the in-plane stresses change linearly throughout a lamina, equation (6.69), but discontinuously from lamina to lamina. For that reason the calculations start with the linearly changing in-plane strain distribution.

The principal directions in the midplane of the kth lamina are $(x_1)_k$ and $(x_2)_k$. The axes in that plane parallel to the principal directions of the laminate, x and y, are $(x_1')_k$ and $(x_2')_k$, respectively, see figure 7.3. The angle between $(x_1)_k$- and $(x_1')_k$-axis is orientation angle θ_k.

Suppose the reduced stiffness matrix of the kth (orthotropic) lamina with respect to the principal directions $(x_1)_k$ and $(x_2)_k$ is $[Q]_k$. Transformation of $[Q]_k$, equation (6.54), gives the reduced stiffness matrix with respect to the principal directions of the laminate, $[Q']_k$. Written in full, without subscript k:

$$Q_{11}' = Q_{11} \cos^4\theta + 2(Q_{12} + 2Q_{66}) \sin^2\theta \cos^2\theta + Q_{22} \sin^4\theta \ ;$$

$$Q_{22}' = Q_{11} \sin^4\theta + 2(Q_{12} + 2 Q_{66}) \sin^2\theta \cos^2\theta + Q_{22} \cos^4\theta \ ;$$

$$Q_{12}' = (Q_{11} + Q_{22} - 4Q_{66}) \sin^2\theta \cos^2\theta + Q_{12} (\sin^4\theta + \cos^4\theta) \ ;$$

$$Q_{66}' = (Q_{11} + Q_{22} - 2Q_{12} - 2Q_{66}) \sin^2\theta \cos^2\theta + Q_{66} (\sin^4\theta + \cos^4\theta) \ ;$$

$$Q_{16}' = (Q_{11} - Q_{12} - 2Q_{66}) \sin\theta \cos^3\theta - (Q_{22} - Q_{12} - 2Q_{66}) \sin^3\theta \cos\theta \ ;$$

$$Q_{26}' = (Q_{11} - Q_{12} - 2Q_{66}) \sin^3\theta \cos\theta - (Q_{22} - Q_{12} - 2Q_{66}) \sin\theta \cos^3\theta.$$

$$(7.1)$$

Equation (6.33) gives the stress-strain relations of the kth lamina with

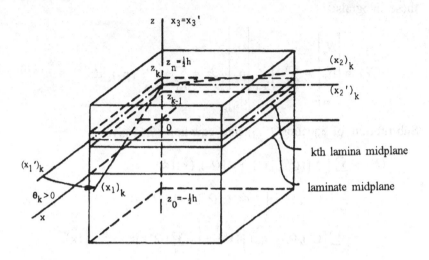

Figure 7.3. The coordinate systems of laminate and laminae.

respect to the principal directions of the lamina as:

$$
\begin{bmatrix} \sigma_1 \\ \sigma_2 \\ \sigma_6 \end{bmatrix}_k = \begin{bmatrix} Q_{11} & Q_{12} & 0 \\ & Q_{22} & 0 \\ & & Q_{66} \end{bmatrix}_k \begin{bmatrix} \varepsilon_1 \\ \varepsilon_2 \\ \varepsilon_6 \end{bmatrix}_k \quad ; \text{ in short}: \{\sigma\}_k = [Q]_k \{\varepsilon\}_k. \tag{7.2}
$$

And with respect to the principal directions of the laminate:

$$
\begin{bmatrix} \sigma_1' \\ \sigma_2' \\ \sigma_6' \end{bmatrix}_k = \begin{bmatrix} Q_{11}' & Q_{12}' & Q_{16}' \\ & Q_{22}' & Q_{26}' \\ & & Q_{66}' \end{bmatrix}_k \begin{bmatrix} \varepsilon_1' \\ \varepsilon_2' \\ \varepsilon_6' \end{bmatrix}_k \quad ; \text{ in short}: \{\sigma'\}_k = [Q']_k \{\varepsilon'\}_k. \tag{7.3}
$$

Equation (6.65) gives the in-plane strain vector $\{\varepsilon'\}_k$ of the kth lamina as a function of z:

$$
\{\varepsilon'\}_k = \{\varepsilon^0\} + z \{\kappa\} ; \quad z_{k-1} < z < z_k . \tag{7.4}
$$

Then the stress-strain relations of the kth lamina can be written as:

$$
\{\sigma'\}_k = [Q']_k \{\varepsilon^0\} + z [Q']_k \{\kappa\} ; \quad z_{k-1} < z < z_k . \tag{7.5}
$$

Because the in-plane stresses show a discontinuity at the interfaces, the integral over the whole thickness in equation (6.67) must be replaced by integrals over the separate laminae and summation of these integrals:

$$\{N\} = \begin{bmatrix} N_x \\ N_y \\ N_{xy} \end{bmatrix} = \sum_{k=1}^{n} \int_{z_{k-1}}^{z_k} \begin{bmatrix} \sigma'_x \\ \sigma'_y \\ \sigma'_{xy} \end{bmatrix}_k dz = \sum_{k=1}^{n} \int_{z_{k-1}}^{z_k} \{\sigma'\}_k \, dz \ . \tag{7.6}$$

Substitution of equation (7.5) into equation (7.6) gives:

$$\begin{aligned} \{N\} &= \sum_{k=1}^{n} \int_{z_{k-1}}^{z_k} \left([Q']_k \{\varepsilon^0\} + z \, [Q']_k \{\kappa\} \right) dz \\ &= \sum_{k=1}^{n} \left([Q']_k \{\varepsilon^0\} \int_{z_{k-1}}^{z_k} dz + [Q']_k \{\kappa\} \int_{z_{k-1}}^{z_k} z \, dz \right) \\ &= \left(\sum_{k=1}^{n} [Q']_k (z_k - z_{k-1}) \right) \{\varepsilon^0\} + \left(\frac{1}{2} \sum_{k=1}^{n} [Q']_k (z_k^2 - z_{k-1}^2) \right) \{\kappa\} \ . \end{aligned} \tag{7.7a}$$

Equation (7.7a) can be written as:

$$\begin{bmatrix} N_x \\ N_y \\ N_{xy} \end{bmatrix} = \begin{bmatrix} A_{11} & A_{12} & A_{16} \\ A_{21} & A_{22} & A_{26} \\ A_{61} & A_{62} & A_{66} \end{bmatrix} \begin{bmatrix} \varepsilon_x^0 \\ \varepsilon_y^0 \\ \varepsilon_{xy}^0 \end{bmatrix} + \begin{bmatrix} B_{11} & B_{12} & B_{16} \\ B_{21} & B_{22} & B_{26} \\ B_{61} & B_{62} & B_{66} \end{bmatrix} \begin{bmatrix} \kappa_x \\ \kappa_y \\ \kappa_{xy} \end{bmatrix} ; \tag{7.7b}$$

or, in short:

$$\{N\} = [A] \{\varepsilon^0\} + [B] \{\kappa\} \ . \tag{7.7c}$$

The elements of the matrices [A] and [B] can be calculated if the laminate construction is known:

$$A_{ij} = \sum_{k=1}^{n} [Q'_{ij}]_k (z_k - z_{k-1}) \ ; \tag{7.8a}$$

$$B_{ij} = \frac{1}{2} \sum_{k=1}^{n} [Q'_{ij}]_k (z_k^2 - z_{k-1}^2) \ ; \qquad i, j \in \{1, 2, 6\} \ . \tag{7.8b}$$

Because the $[Q']_k$'s are symmetric matrices, the matrices [A] and [B] are also symmetric:

$$A_{ji} = A_{ij} \ ; \quad B_{ji} = B_{ij} \ . \tag{7.9a}$$

Analogous to equation (7.6) the moments per unit length of edge from equation (6.68) are now expressed as a summation:

$$\{M\} = \begin{bmatrix} M_x \\ M_y \\ M_{xy} \end{bmatrix} = \sum_{k=1}^{n} \int_{z_{k-1}}^{z_k} \begin{bmatrix} \sigma'_x \\ \sigma'_y \\ \sigma'_{xy} \end{bmatrix}_k z \, dz = \sum_{k=1}^{n} \int_{z_{k-1}}^{z_k} \{\sigma'\}_k \, z \, dz \; . \quad (7.10)$$

Substitution of equation (7.5) into equation (7.10) gives a result analogous to equation (7.7b):

$$\begin{bmatrix} M_x \\ M_y \\ M_{xy} \end{bmatrix} = \begin{bmatrix} B_{11} & B_{12} & B_{16} \\ B_{21} & B_{22} & B_{26} \\ B_{61} & B_{62} & B_{66} \end{bmatrix} \begin{bmatrix} \varepsilon^0_x \\ \varepsilon^0_y \\ \varepsilon^0_{xy} \end{bmatrix} + \begin{bmatrix} D_{11} & D_{12} & D_{16} \\ D_{21} & D_{22} & D_{26} \\ D_{61} & D_{62} & D_{66} \end{bmatrix} \begin{bmatrix} \kappa_x \\ \kappa_y \\ \kappa_{xy} \end{bmatrix} \; ; \qquad (7.11a)$$

in short:

$$\{M\} = [B] \{\varepsilon^0\} + [D] \{\kappa\} \; . \qquad (7.11b)$$

For the elements of [B] equation (7.8b) holds, and for the elements of [D]:

$$D_{ij} = \frac{1}{3} \sum_{k=1}^{n} [Q'_{ij}]_k \, (z_k^3 - z_{k-1}^3) \; ; \quad i,j \in \{1,2,6\} \; . \qquad (7.8c)$$

Matrix [D] is also symmetric:

$$D_{ji} = D_{ij} \; . \qquad (7.9b)$$

The relations (7.7) and (7.11) can be combined as being the load-strain relations of the laminate according to the classical laminate theory (CLT):

$$\begin{bmatrix} N_x \\ N_y \\ N_{xy} \\ \dots \\ M_x \\ M_y \\ M_{xy} \end{bmatrix} = \begin{bmatrix} A_{11} & A_{12} & A_{16} & \vdots & B_{11} & B_{12} & B_{16} \\ A_{21} & A_{22} & A_{26} & \vdots & B_{21} & B_{22} & B_{26} \\ A_{61} & A_{62} & A_{66} & \vdots & B_{61} & B_{62} & B_{66} \\ \dots & \dots & \dots & \dots & \dots & \dots & \dots \\ B_{11} & B_{12} & B_{16} & \vdots & D_{11} & D_{12} & D_{16} \\ B_{21} & B_{22} & B_{26} & \vdots & D_{21} & D_{22} & D_{26} \\ B_{61} & B_{62} & B_{66} & \vdots & D_{61} & D_{62} & D_{66} \end{bmatrix} \begin{bmatrix} \varepsilon^0_x \\ \varepsilon^0_y \\ \varepsilon^0_{xy} \\ \dots \\ \kappa_x \\ \kappa_y \\ \kappa_{xy} \end{bmatrix} \; ; \qquad (7.12a)$$

in short:

$$
\begin{bmatrix} \{N\} \\ .... \\ \{M\} \end{bmatrix} = \begin{bmatrix} [A] & \vdots & [B] \\ .... & .... & .... \\ [B] & \vdots & [D] \end{bmatrix} \begin{bmatrix} \{\varepsilon^0\} \\ .... \\ \{\kappa\} \end{bmatrix} . \tag{7.12b}
$$

Because [A], [B] and [D] are symmetric, the 6x6-matrix is also symmetric. The elements of [A], [B] and [D] are stiffness constants.

Matrix [A] determines the in-plane stiffness of the laminate under influence of tension, compression and shearing. The elements A_{ij} are called the in-plane stiffnesses of the laminate. They are independent on the stacking sequence of the laminae. The weighting factor is the thickness of the lamina, z_k-z_{k-1}.

Matrix [D] determines the stiffness of the laminate in a perpendicular direction under influence of bending and twisting moments. The elements D_{ij} are the flexural stiffnesses of the laminate. They are strongly dependent on the sequence of the laminae.

Matrix [B] is responsible for coupling effects, on the one hand between in-plane stresses and curvatures, on the other hand between bending and twisting moments and in-plane strains. Matrix [B] is called the coupling matrix, and the elements B_{ij} are coupling stiffnesses, dependent on the stacking sequence.

The stiffnesses A_{ij}, B_{ij} and D_{ij} do not all have the same dimensions. If F symbolises the dimension "force", and L the dimension "length", then the different quantities will have the following dimensions:

$$\{\sigma\} \text{ in } [FL^{-2}] ; \quad \{\varepsilon\} \text{ in } [1] ; \quad \{\kappa\} \text{ in } [L^{-1}] ;$$

$$\{Q\} \text{ in } [FL^{-2}] ; \quad \{N\} \text{ in } [FL^{-1}] ; \quad \{M\} \text{ in } [F] ;$$

$$[A] \text{ in } [FL^{-1}] ; \quad [B] \text{ in } [F] ; \quad [D] \text{ in } [FL] .$$

Generally the laminate stiffnesses are dependent on the orientation. In the case of rotation of the coordinate system around a normal line of the midplane the same transformation equation can be used as for [Q], namely equation (6.53a).

$$
\begin{aligned}
[A''] &= [T] \, [A] \, [T^T] ; \\
[B''] &= [T] \, [B] \, [T^T] ; \\
[D''] &= [T] \, [D] \, [T^T] ;
\end{aligned}
\tag{7.13}
$$

with [T] equation (6.48a) expressed in the positive transformation angle φ. A double prime is used to avoid confusion with the transformations with respect to orientation angle θ.

7.3 DETERMINATION OF LAMINAE STRESSES AND STRAINS

All the ingredients of the recipe have now been assembled. The stiffness analysis of a laminate can start. Suppose the laminate is made up of known orthotropic laminae, and the sequence and orientation of the laminae - rank number k and angle θ_k, respectively - are given.   The stiffness constants $[Q]_k$ of the kth lamina with respect to its principal directions are known. Equation (7.1) produces the values of the stiffness constants with respect to the principal directions of the laminate, $[Q']_k$.

Now the stiffness matrices [A], [B] and [D] can be calculated with help of equation (7.8). Suppose the loads [N] and [M] are imposed with respect to the principal directions of the laminate. To calculate the deformations of the midplane of the laminate it is necessary to invert equation (7.12). The result is:

$$
\begin{bmatrix} \{\varepsilon^0\} \\ \cdots \\ \{\kappa\} \end{bmatrix} = \begin{bmatrix} [a] & \vdots & [b] \\ \cdots & \cdots & \cdots \\ [b^T] & \vdots & [d] \end{bmatrix} \begin{bmatrix} \{N\} \\ \cdots \\ \{M\} \end{bmatrix} . \tag{7.14}
$$

The matrices [a], [b] and [d] are compliance matrices with dimensions equal to the reciprocal of the dimensions of the respective stiffness matrices [A], [B] and [D]. Matrix [a] is the in-plane compliance matrix, [b] is the coupling compliance matrix, and [d] is the flexural compliance matrix. The whole 6x6-matrix is symmetric, but [b] is not symmetric. Inversion after partitioning results in the relations

$$ [a] = [A]^{-1} - [b] [B] [A]^{-1} ; $$

$$ [b] = -[A]^{-1} [B] [d] ; \tag{7.15} $$

$$ [d] = ([D] - [B] [A]^{-1} [B])^{-1} . $$

Equation (7.4) brings the in-plane strains $[\varepsilon']_k$ with respect to the principal directions of the laminate. With help of the inverse of equation (6.51) the in-plane strains $[\varepsilon]_k$ with respect to the principal directions of the kth lamina can be calculated.

$$
\begin{bmatrix} \varepsilon_1 \\ \varepsilon_2 \\ \varepsilon_6 \end{bmatrix}_k = \begin{bmatrix} \cos^2\theta & \sin^2\theta & \sin\theta\,\cos\theta \\ \sin^2\theta & \cos^2\theta & -\sin\theta\,\cos\theta \\ -2\sin\theta\,\cos\theta & 2\sin\theta\,\cos\theta & \cos^2\theta - \sin^2\theta \end{bmatrix}_k \begin{bmatrix} \varepsilon_1' \\ \varepsilon_2' \\ \varepsilon_6' \end{bmatrix}_k . \tag{7.16}
$$

Then Hooke's law, equation (6.33), gives the in-plane stresses $[\sigma]_k$.

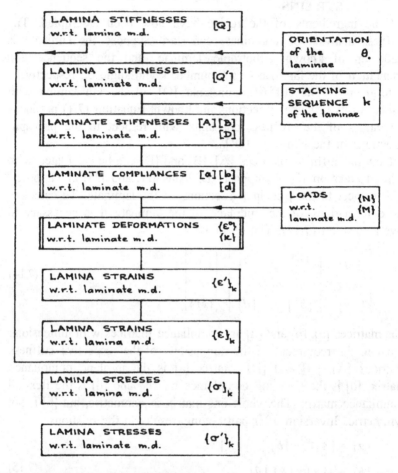

Figure 7.4. Flow chart for stiffness analysis of a laminate, and the calculation of the lamina stresses and strains. Quantities are given with respect to the principal (or main) directions of lamina or laminate.

If desired the in-plane stresses $[\sigma']_k$ with respect to the principal directions of the laminate can be calculated by transformation:

$$
\begin{bmatrix} \sigma'_1 \\ \sigma'_2 \\ \sigma'_6 \end{bmatrix}_k = \begin{bmatrix} \cos^2\theta & \sin^2\theta & -2\sin\theta\,\cos\theta \\ \sin^2\theta & \cos^2\theta & 2\sin\theta\,\cos\theta \\ \sin\theta\,\cos\theta & -\sin\theta\,\cos\theta & \cos^2\theta-\sin^2\theta \end{bmatrix}_k \begin{bmatrix} \sigma_1 \\ \sigma_2 \\ \sigma_6 \end{bmatrix}_k . \qquad (7.17)
$$

The flow chart in figure 7.4 demonstrates the stiffness analysis of the laminate.

To analyse the strength of the laminate it is necessary to know the laminae stresses or strains. The usual fracture criteria of orthotropic laminae are expressed in either stresses or strains, and with respect to the principal directions of the lamina. As soon as these laminae stresses or strains are known the chosen fracture criterion can be applied. In figure 7.5 the flowpath for strength analysis is indicated.

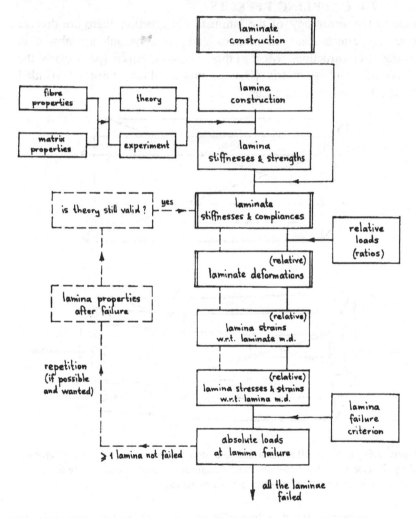

Figure 7.5. Flow chart for strength analysis of a laminate.

Generally it is the best to analyse every lamina. There are many PC programs available to do it. The point at which the first lamina fractures is called first ply failure (FPF). In some cases the laminate can still fulfil its function. Then it is possible to introduce the properties of the failed lamina and to calculate the behaviour of the laminate with the broken lamina. For instance this could be in the case of cracks in the matrix of a lamina in the wall of a vessel with a thermoplastic lining.

7.4 COUPLING EFFECTS

Due to the anisotropy of the laminate construction there are diverse coupling effects between stresses and strains which are absent in isotropic constructions. Most of these coupling effects (particularly the effect of coupling matrix [B]) are unwanted and must be avoided except for very special cases.

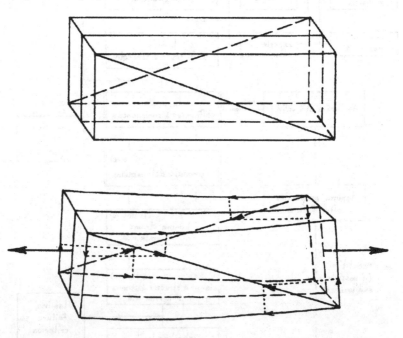

Figure 7.6. Twisting effect in a laminate strip caused by a tensile force directed along the bisector of the respective fibre directions of the two laminae. The generated tensile forces in the fibres produce a twisting moment.

To demonstrate the coupling effects it is easier to look primarily at the coupling compliance matrix [b]. Due to coupling compliance b_{11} a uniaxial load N_x causes a curvature κ_x of the x-axis, and a bending moment M_x causes a strain ε_x. Due to coupling compliance b_{61} N_x

causes a twisting curvature κ_{xy}, see figure 7.6, and M_x causes a shear strain ε_{xy}, etc. To avoid these coupling effects it is sufficient to make [b]=[0], or [B]=[0] as equation (7.15) demonstrates. From equation (7.8b):

$$B_{ij} = \frac{1}{2} \sum_{k=1}^{n} [Q'_{ij}]_k \, (z_k^2 - z_{k-1}^2) \; ; \quad i,j \in \{1,2,6\} \, . \tag{7.8b}$$

The simplest solution to avoid the main coupling effects due to [B] is to construct the laminate symmetrically with respect to the midplane.

In that case the whole number of laminae can be divided in pairs in such a way that for each pair one lamina is the image of the other with respect to the midplane, see figure 7.7.

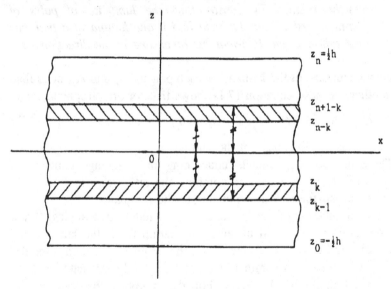

Figure 7.7. Symmetrical position of the kth and the (n-1+k)th laminae with respect to the laminate midplane. The orientation of both identical laminae must be equal.

The kth lamina constitutes a pair with the (n+1-k)th lamina. Both laminae of each pair are identical and have the same orientation:

$$[Q'_{ij}]_k = [Q'_{ij}]_{n+1-k} \, .$$

But:

$$z_k^2 - z_{k-1}^2 = - (z_{n+1-k}^2 - z_{n-k}^2) .$$

Substitution in equation (7.8b) shows that the contributions of both laminae to B_{ij} neutralise each other. And so do all pairs of laminae.

Further, it is possible to make the laminate orthotropic in relation to the in-plane stiffness matrix [A], that means to realise $A_{16}=A_{26}=0$ ($=A_{61}=A_{62}$) with respect to the principal directions of the laminate. The remaining in-plane stiffnesses are always positive, otherwise the material would not have the ability to bear load. From equation (7.8a):

$$A_{ij} = \sum_{k=1}^{n} [Q_{ij}']_k (z_k - z_{k-1}). \qquad (7.8a)$$

shows that

> Orthotropy in relation to the in-plane stiffness of the laminate can be created by constructing the laminate of pairs of identical orthotropic laminae and if one lamina of a pair has orientation angle θ_i orient the other one in the direction $-\theta_i$.

The contributions of the laminae of each pair to A_{16} and A_{26} neutralise each other, because equation (7.1) shows that for orthotropic laminae:

$$Q_{16}' (-\theta_i) = - Q_{16}' (\theta_i) ; \quad Q_{26}' (-\theta_i) = - Q_{26}' (\theta_i) , \qquad (7.18)$$

and z_k-z_{k-1} is the thickness of the lamina.

There are no requirements concerning the stacking sequence of these laminae. So it is easy to combine the orthotropy in relation to the in-plane stiffness with symmetry to the midplane.

If the orientation of the laminae is restricted to two angles, θ and $-\theta$, the laminate is called an angle-ply laminate. If the laminae are identical and lie in an alternate order, the laminate is called a regular angle-ply laminate. A symmetric regular angle-ply laminate is called a balanced angle-ply laminate. For that purpose the two middle laminae should have the same orientation.

Laminae with orientation angle $0°$ or $90°$ make no contribution to A_{16} and A_{26}. They can be added to an angle-ply laminate without disturbing the orthotropy. A laminate of laminae with orientations of $0°$ and $90°$ only is called a cross-ply laminate. A cross-ply laminate is always orthotropic in relation to the in-plane stiffness.

Orthotropy in relation to [D], that means $D_{16}=D_{26}=0$, is also possible. From equation (7.8c):

$$D_{ij} = \frac{1}{3} \sum_{k=1}^{n} [Q'_{ij}]_k (z_k^3 - z_{k-1}^3) ; \quad i,j \in \{1,2,6\} . \qquad (7.8c)$$

It can be seen that:

Orthotropy in relation to [D] can be realised by constructing the laminate antisymmetrically with respect to the midplane.

That means that every lamina with orientation θ_i is combined with an identical lamina that lies at the same distance of the midplane on the other side and has orientation $-\theta_i$, see figure 7.8. The requirement of an antisymmetric construction is in contradiction to the requirement of a symmetric construction in order to make [B]=[0]. In practice the last requirement (on [B]) is more imperative.

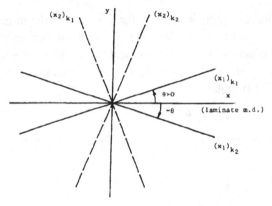

Figure 7.8.   Two laminae (the k_1th and the k_2th) under an angle θ and $-\theta$, respectively, with the x-axis of the laminate.

Some special constructions are uncoupled - [B]=[0] - and orthotropic in relation to [A] and [B]. Two examples are taken from Bartholomew [1]. Firstly, an antisymmetric angle-ply laminate of 8 laminae:

$$[\theta/-\theta/-\theta/\theta/-\theta/\theta/\theta/-\theta]. \qquad (7.19)$$

Secondly, a symmetric laminate of 12 identical laminae as a combination from angle-plies and laminae in 0° and/or 90° direction, complying with symmetry. This choice is indicated by 0|90.

$$[\theta / -\theta / -\theta / 0|90 / \theta / 0|90]_s . \qquad (7.20)$$

7.5 CONSTRUCTION AND PROPERTIES OF SPECIAL LAMINATES

7.5.1 Symmetric Laminates

In the previous section it is mentioned that the coupling effects due to [B] could be avoided by constructing the laminate symmetrically in relation to its midplane. In that case it is meaningful to introduce in-plane engineering constants of the laminate, because a uniaxial in-plane force, e.g. N_x, will not cause bending or twisting. The starting point is a global in-plane stress distribution, averaged across the laminate thickness. Therefore the forces per unit length of an edge are divided by thickness h.

$$\overline{\sigma}_x = \frac{N_x}{h} \; ; \quad \overline{\sigma}_y = \frac{N_y}{h} \; ; \quad \overline{\sigma}_{xy} = \frac{N_{xy}}{h} \; . \tag{7.21}$$

If a uniaxial force N_x acts in the laminate, generally the in-plane stress distribution in the respective laminae is not a uniaxial one.

Suppose the only non-zero component of the loading vector is N_x. According to equation (7.14) the strain ε_x is constant and equal to the strain ε_x^0 in the midplane.

$$\varepsilon_x^0 = a_{11} \, N_x = a_{11} \, \overline{\sigma}_x \, h \; .$$

The modulus of elasticity, E_x^0, of the laminate can be defined as follows:

$$E_x^0 = \frac{\overline{\sigma}_x}{\varepsilon_x^0} = \frac{1}{a_{11} h} \; . \tag{7.22}$$

The cross strain of the midplane in y-direction caused by N_x is:

$$\varepsilon_y^{0(x)} = a_{21} \, N_x = a_{21} \, \overline{\sigma}_x \, h.$$

With this the Poisson's ratio v_{xy} is given:

$$v_{xy} = - \frac{\varepsilon_y^{0(x)}}{\varepsilon_x^0} = - \frac{a_{21}}{a_{11}} \; . \tag{7.23}$$

In the same way as E_x^0 the modulus of elasticity in y-direction, E_y^0, can be expressed in terms of an in-plane compliance component:

$$E_y^0 = \frac{\overline{\sigma}_y}{\varepsilon_y^0} = \frac{1}{a_{22} h} \; . \tag{7.24}$$

Because of reciprocity:

$$\frac{v^0_{yx}}{E^0_y} = \frac{v^0_{xy}}{E^0_x} . \qquad (7.25)$$

Finally, if only N_{xy} acts, the shear strain ε_{xy} is constant and equal to $\varepsilon_{xy}{}^0$:

$$\varepsilon^0_{xy} = a_{66} N_{xy} = a_{66} \overline{\sigma}_{xy} h .$$

Then the shear modulus E^0_s of the laminate is by definition equal to:

$$E^0_s \ (= G^0_{xy}) = \frac{\overline{\sigma}_{xy}}{\varepsilon^0_{xy}} = \frac{1}{a_{66} h} . \qquad (7.26)$$

In the case of loadings applied in directions other than principal directions or of non-orthotropy it is difficult to determine the engineering constants of the laminate by experiments. A tensile test on a strip will cause shearing between tensile direction and cross-wise direction. Avoiding this shearing effect by fixed clamps introduces unwanted stresses and disturbs the uniaxial loading condition. The clamps should be provided with a mechanism to allow rotation around the centre of the clamping line.

7.5.2 Orthotropic Laminates

Often a laminated object has fixed in-plane loading directions. Then an orthotropic laminate with principal directions corresponding to the loading directions is preferable, resulting in a symmetric deformation. Symmetry with respect to the midplane and orthotropy in relation to [A] are required.

With respect to the axes of symmetry:

$$A_{16} = A_{26} = 0 .$$

The in-plane compliances are then according to equation (7.15):

$$a_{11} = \frac{A_{22}}{A_{11}A_{22} - A^2_{12}} ; \quad a_{22} = \frac{A_{11}}{A_{11}A_{22} - A^2_{12}} ;$$

$$a_{12} = -\frac{A_{12}}{A_{11}A_{22} - A^2_{12}} = a_{21} = -\frac{A_{21}}{A_{11}A_{22} - A^2_{12}} ; \qquad (7.27)$$

$$a_{16} = a_{26} = 0; \quad a_{66} = \frac{1}{A_{66}} .$$

Substitution of equation (7.27) into equations (7.22), (7.23), (7.24) and (7.26) leads to the expression for the in-plane engineering constants in the in-plane stiffnesses:

$$E_x^0 = \frac{A_{11}A_{22} - A_{12}^2}{hA_{22}} \; ; \quad E_y^0 = \frac{A_{11}A_{22} - A_{12}^2}{hA_{11}} \; ;$$

$$v_{xy}^0 = \frac{A_{12}}{A_{22}} \; ; \qquad G_{xy}^0 = \frac{A_{66}}{h} \; .$$

(7.28)

An example of an orthotropic laminate is a balanced angle-ply laminate, consisting of identical orthotropic laminae. Fraction f_θ of the whole number of laminae is oriented in θ-direction and a same fraction $f_{-\theta}$ in $-\theta$-direction. The angle-plies are completed with a fraction f_0 in $0°$-direction and a fraction f_{90} in $90°$-direction.

$$f_0 + f_\theta + f_{-\theta} + f_{90} = 1 \; ; \quad f_{-\theta} = f_\theta \; .$$

(7.29)

According to equation (7.8a) the in-plane stiffnesses are:

$$A_{ij} = h \{ f_0 \cdot Q_{ij} + f_\theta \cdot Q_{ij}'(\theta) + f_\theta \cdot Q_{ij}'(-\theta) + f_{90} \cdot Q_{ij}'(90) \} .$$

(7.30)

Substituting equation (7.30) into equation (7.28) the in-plane engineering constants are expressed in terms of the reduced stiffnesses of the laminae as:

$$E_x^0 = f_0 \cdot Q_{11} + 2 f_\theta \cdot Q_{11}'(\theta) + f_{90} \cdot Q_{22} - \frac{\{ (1 - 2 f_\theta) \cdot Q_{12} + 2 f_\theta \cdot Q_{12}'(\theta) \}^2}{f_0 \cdot Q_{22} + 2 f_\theta \cdot Q_{22}'(\theta) + f_{90} \cdot Q_{11}} \; ;$$

$$E_y^0 = f_0 \cdot Q_{22} + 2 f_\theta \cdot Q_{22}'(\theta) + f_{90} \cdot Q_{11} - \frac{\{ (1 - 2 f_\theta) \cdot Q_{12} + 2 f_\theta \cdot Q_{12}'(\theta) \}^2}{f_0 \cdot Q_{11} + 2 f_\theta \cdot Q_{11}'(\theta) + f_{90} \cdot Q_{22}} \; ;$$

$$v_{xy}^0 = \frac{(1 - 2 f_\theta) \cdot Q_{12} + 2 f_\theta \cdot Q_{12}'(\theta)}{f_0 \cdot Q_{22} + 2 f_\theta \cdot Q_{22}'(\theta) + f_{90} \cdot Q_{11}} \; ;$$

$$G_{xy}^0 = (1 - 2 f_\theta) \cdot Q_{66} + 2 f_\theta \cdot Q_{66}'(\theta) .$$

(7.31)

Substitution of equations (6.34) and (7.1) into equation (7.31) leads to relations between the in-plane engineering constants of the total balanced angle-ply laminate and the (in-plane) engineering constants of the individual laminae. The details are omitted here.

If it is required to optimise the laminate, it is recommended that various geometries be analysed. For instance, if a certain ratio of E_x and E_y is required, calculations will demonstrate that a cross-ply laminate can be thinner than an angle-ply laminate with equal main in-plane extensional stiffnesses A_{11} and A_{22}.

7.5.3 Plane-Isotropic Laminates

Sometimes, for reason of chemical durability, for example, it is desirable to apply FRP without utilising the advantages of anisotropy. A laminate construction of laminae reinforced with non-woven fabrics in which the fibres are at random orientated in a plane can be used as example. Such a laminate shows isotropy in its plane, here indicated with the term plane-isotropy.

To predict the engineering constants of a plane-isotropic laminate the following model is useful. The laminate is composed of many identical unidirectional laminae of infinitely small thickness. Then the summations in equation (7.8) can be changed by integrations:

$$A_{ij} = \int_{-\frac{1}{2}h}^{\frac{1}{2}h} Q'_{ij} \, dz \; ;$$

$$B_{ij} = \int_{-\frac{1}{2}h}^{\frac{1}{2}h} Q'_{ij} \, z \, dz \; ; \qquad\qquad (7.32)$$

$$D_{ij} = \int_{-\frac{1}{2}h}^{\frac{1}{2}h} Q'_{ij} \, z^2 \, dz \; .$$

The reduced stiffnesses Q_{ij}' are functions of orientation angle θ according to equation (7.1). Suppose the sequence of the laminae is arbitrary. Then angle θ is independent to coordinate z. Angle θ can take all values between $-\pi/2$ and $\pi/2$ with the same probability. For that reason equation (7.32) can be rewritten as:

$$A_{ij} = \frac{1}{\pi} \int_{-\frac{1}{2}\pi}^{\frac{1}{2}\pi} Q'_{ij} \, d\theta \int_{-\frac{1}{2}h}^{\frac{1}{2}h} dz \; ;$$

$$B_{ij} = \frac{1}{\pi} \int_{-\frac{1}{2}\pi}^{\frac{1}{2}\pi} Q'_{ij} \, d\theta \int_{-\frac{1}{2}h}^{\frac{1}{2}h} z \, dz \; ; \qquad\qquad (7.33)$$

$$D_{ij} = \frac{1}{\pi} \int_{-\frac{1}{2}\pi}^{\frac{1}{2}\pi} Q'_{ij} \, d\theta \int_{-\frac{1}{2}h}^{\frac{1}{2}h} z^2 \, dz \; .$$

Now:

$$\int_{-\frac{1}{2}\pi}^{\frac{1}{2}\pi} \cos 2p\theta \, d\theta = \int_{-\frac{1}{2}\pi}^{\frac{1}{2}\pi} \sin 2p\theta \, d\theta = 0 \quad \text{for } p = 1,2,\dots \quad (7.34)$$

Substitution of equation (6.56) into equation (7.33) and using equation (7.34) results in:

$$A_{11} = A_{22} = hU_1 \; ;$$

$$A_{12} = hU_4 \; ;$$

$$A_{66} = hU_5 \; ; \qquad\qquad\qquad\qquad (7.35)$$

$$A_{16} = A_{26} = 0 \; .$$

$$B_{ij} = 0 \; . \qquad\qquad\qquad\qquad\qquad (7.36)$$

$$D_{ij} = \frac{1}{12} h^2 A_{ij} \; . \qquad\qquad\qquad\qquad (7.37)$$

Because U_1, U_4, and U_5 are invariant, A_{ij} and D_{ij} are also invariant. The laminate behaves isotropically in relation to both [A] and [D], while [B] has disappeared. Introduction of the relation between U_1, U_4, and U_5 in equation (6.59) into equation (7.35) gives:

$$A_{66} = \frac{1}{2} (A_{11} - A_{12}) \; . \qquad\qquad\qquad (7.38)$$

There remain two independent elastic constants which describe the elastic behaviour of the plane-isotropic laminate in accordance with the classical laminate theory including the basic assumptions. This behaviour is like that of an isotropic plate in the same circumstances.

Substitution of equation (7.35) into equation (7.28) brings the in-plane engineering constants of the plane-isotropic laminate to the form:

$$E = U_1 - \frac{U_4^2}{U_1} \; . \qquad\qquad\qquad\qquad (7.39)$$

$$\bar{v} = \frac{U_4}{U_1} \; . \qquad\qquad\qquad\qquad\qquad (7.40)$$

$$\overline{G} = U_5 . \tag{7.41}$$

As was only to be expected, equations (7.39), (7.40) and (7.41) show the well-known interrelation of the engineering constants of isotropic materials:

$$\overline{G} = \frac{E}{2\,(1+\overline{\nu})} . \tag{7.42}$$

If the engineering constants of the unidirectional lamina are known, the reduced stiffnesses can be calculated with help of equation (6.34).

Substitution of the results in equation (6.57) gives:

$$U_1 = \frac{3}{8} \, \frac{1}{1-\nu_{LT}\nu_{TL}} \, \{E_L + (1+\tfrac{2}{3}\,\nu_{LT})\,E_T\} + \tfrac{1}{2}\,G_{LT} \, ;$$

$$U_4 = \frac{1}{8} \, \frac{1}{1-\nu_{LT}\nu_{TL}} \, \{E_L + (1+6\nu_{LT})\,E_T\} - \tfrac{1}{2}\,G_{LT} \, ; \tag{7.43}$$

$$U_5 = \frac{1}{8} \, \frac{1}{1-\nu_{LT}\nu_{TL}} \, \{E_L + (1-2\nu_{LT})\,E_T\} + \tfrac{1}{2}\,G_{LT} \, .$$

And now the engineering constants in equations (7.39), (7.40) and (7.41) can be expressed in terms of the engineering constants of a unidirectional lamina.

7.5.4 Quasi-Isotropic and Quasi-Homogeneous Laminates

In the previous section, plane-isotropic laminates built of laminae with randomly distributed fibres, were considered. It is also possible to make laminates of identical orthotropic laminae which have an isotropic stiffness behaviour in their plane, the so-called quasi-isotropic laminates. Generally these laminates are not completely isotropic and it is necessary to specify which elastic properties of the laminate appear isotropic.

There can be quasi-isotropy in relation to the in-plane stiffnesses [A], the so-called in-plane quasi-isotropy. Then the following conditions are satisfied:

$$A_{11} = A_{22} \, ; \quad A_{66} = \tfrac{1}{2}\,(A_{11} - A_{12}) \, ; \quad A_{16} = A_{26} = 0 \, ; \tag{7.44}$$

securing direction independency in the xy-plane.

Suppose the laminate has n identical orthotropic laminae; the lamina thickness is h/n. According to equation (7.8a):

$$A_{ij} = \frac{h}{n} \sum_{k=1}^{n} \left(Q'_{ij} \right)_k ,$$ (7.45)

irrespective of the stacking sequence.

The number of laminae can be divided in r groups each with the same number of laminae, i.e., n/r. The laminae within a group have the same orientation angle, θ_m. The angle between adjacent directions is constant, that is π/r. If the x-axis is directed in the principal direction of one group, the different orientation angles are:

$$\theta_m = m \frac{\pi}{r} ; \quad m \in \{1,...,r\} ; \quad r \geq 3 .$$ (7.46)

In this manner $\theta_r = \pi$ also corresponds to the x-axis. Later, it will be seen that the number of different directions should be at least three. Once again, the stacking sequence of all the laminae is unrestricted. Figure 7.9 gives an example of systematically distributed orientations to give in-plane quasi-isotropy for the laminate.

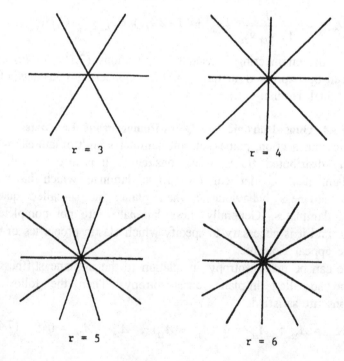

Figure 7.9. Examples of systematically distributed orientations (number $r \geq 3$) of the laminae to attain in-plane quasi-isotropy of the laminate.

Substitution of equation (6.56) into equation (7.45) the in-plane stiffnesses A_{ij} are expressed in terms of $U_1,..., U_5$, and sine and cosine of the double and the quadruple of the orientation angle. For instance:

$$A_{11} = \frac{h}{n}\left[U_1 \sum_{k=1}^{n} 1 + U_2 \sum_{k=1}^{n} \cos 2\theta_k + U_3 \sum_{k=1}^{n} \cos 4\theta_k \right] =$$

$$= \frac{h}{n}\left[U_1.n + U_2.\frac{n}{r} \sum_{m=1}^{r} \cos 2\theta_m + U_3.\frac{n}{r} \sum_{m=1}^{r} \cos 4\theta_m \right] =$$

$$= hU_1 + hU_2.\frac{1}{r} \sum_{m=1}^{r} \cos 2\theta_m + hU_3.\frac{1}{r} \sum_{m=1}^{r} \cos 4\theta_m .$$

Now:

$$\sum_{m=1}^{r} \cos 2m\frac{\pi}{r} = \sum_{m=1}^{r} \sin 2m\frac{\pi}{r} = 0 \quad for\ r \geq 2 ; \quad (7.47)$$

$$\sum_{m=1}^{r} \cos 4m\frac{\pi}{r} = \sum_{m=1}^{r} \sin 4m\frac{\pi}{r} = 0 \quad for\ r \geq 3 . \quad (7.48)$$

Substitution of equations (7.46), (7.47) and (7.48) into the expression for A_{11} gives, after some reduction, and under the condition $r \geq 3$:

$$A_{11} = hU_1 .$$

The complete result is the same as in equation (7.35) and so the conditions of equation (7.44) are satisfied. Generally an in-plane quasi-isotropic laminate is not quasi-isotropic in off-plane loading situations.

It is, however, possible to create laminates which are quasi-isotropic in relation to the flexural stiffnesses [D]. Then:

$$D_{11} = D_{22} ; \quad D_{66} = \frac{1}{2} (D_{11} - D_{12}) ; \quad D_{16} = D_{26} = 0 \quad (7.49)$$

and this is called flexural quasi-isotropy.

A laminate that is quasi-isotropic in relation to both [A] and [D], and also is uncoupled ([B] = [0]) behaves with respect to stiffness like an isotropic plate. Then the laminate is said to be quasi-homogeneous. Then, as for isotropic materials:

$$D_{ij} = \frac{1}{12} h^2 A_{ij} \; ; \quad i,j \in \{1,2,6\} \; .$$ (7.50)

The most simple example of a quasi-homogeneous laminate has 18 laminae and three orientations:

$$[0/-60/60/60/-60/0/-60/60/0/0/60/-60/60/0/-60/-60/0/60] .$$

(7.51)

The stacking sequence is not symmetric, although the laminate is uncoupled. Symmetry need not be a condition for uncoupling but it is the simplest way to realise it.

7.6 HYGROTHERMAL STRESSES IN LAMINATES

So far attention is paid to stresses and strains in the laminate and in parts of it caused by external mechanical loads. But even in a stationary state, stresses and strains can be created by other causes: a change of temperature or of moisture content. FRP's also behave anisotropically in relation to these effects.

Generally the coefficient of thermal expansion of the matrix material differs from that of the fibres. And so the linear coefficient of thermal expansion of FRP depends on the direction of measurement. Further, the capabilities of fibres and matrix to absorb moisture differ from each other. Absorbing moisture leads to swelling, and the rate of swelling depends on the direction of measurement, too.

Deformations developed in this way cause no global stresses, because there is equilibrium and no external loads are involved. But, due to the difference in properties of fibres and matrix, these deformations in many cases cause internal stresses which influence the mechanical properties of the composite material. Both effects - thermal expansion and swelling by moisture absorption - are jointly referred to as hygrothermal behaviour.

The theoretical attack of the problem will be explained for thermal behaviour. Suppose the FRP having a temperature T_0 is without stresses. After changing the temperature to T the material is thermally expanded. Usually the assumption can be made that the deformations are proportional to the difference in temperature

$$\Delta T = T - T_0 \; .$$ (7.52)

The deformed state of the material can be described by a strain tensor ε_{ij}^t, with superscript t to designate the thermal cause, by analogy with equation (6.2a).

The proportionality constants between ε_{ij}^t and ΔT are called the thermal expansion coefficients. They are dependent on the orientation and are designated by α_{ij} in uncontracted notation:

$$
\begin{bmatrix} \varepsilon_{11}^t & \varepsilon_{12}^t & \varepsilon_{13}^t \\ & \varepsilon_{22}^t & \varepsilon_{23}^t \\ & & \varepsilon_{33}^t \end{bmatrix} = \begin{bmatrix} \alpha_{11} & \alpha_{12} & \alpha_{13} \\ & \alpha_{22} & \alpha_{23} \\ & & \alpha_{33} \end{bmatrix} \Delta T \ . \tag{7.53a}
$$

The thermal expansion tensor α_{ij} is symmetric for the same reason that ε_{ij}^t is symmetric. Transformation formulae for ε_{ij}^t and α_{ij} are of identical form.

In contracted notation the strain vector $\{\varepsilon^t\}$ is introduced, by analogy with equation (6.2b):

$$
\varepsilon_i^t \ ; \quad i \in \{1,...,6\} \ . \tag{7.54}
$$

In this notation equation (7.53a) can be rewritten as:

$$
\varepsilon_i^t = \alpha_i \ \Delta T \ ; \quad i \in \{1,...,6\} \ , \tag{7.53b}
$$

where

$$
\alpha_1 = \alpha_{11} \, , \ \alpha_2 = \alpha_{22} \, , \ \alpha_3 = \alpha_{33} \, , \ \alpha_4 = 2\alpha_{23} \, , \ \alpha_5 = 2\alpha_{31} \, , \ \alpha_6 = 2\alpha_{12} \ .
$$

Notice the definitions of α_4, α_5 and α_6, which are twice the corresponding tensor component. The thermal expansion vector $\{\alpha\}$ in contracted notation can be transformed with the help of the same transformation matrix as used for $\{\varepsilon\}$.

The entire deformations are a summation of the deformations due to stresses, see equation (6.4), and the thermal deformations:

$$
\varepsilon_i = \sum_{j=1}^{6} S_{ij} \ \sigma_j + \alpha_i \ \Delta T \ ; \quad i \in \{1,...,6\} \ . \tag{7.55}
$$

The part of the deformations due to the real stresses is the difference between the entire strains, ε_i, and the strains caused by expansion, ε_i^t.

$$
\varepsilon_i - \varepsilon_i^t = \varepsilon_i - \alpha_i \ \Delta T = \sum_{j=1}^{6} S_{ij} \ \sigma_j \ .
$$

Inversion leads to an expression for the stresses of the form

$$
\sigma_i = \sum_{j=1}^{6} C_{ij} \ (\varepsilon_j - \varepsilon_j^t) = \sum_{j=1}^{6} C_{ij} \ (\varepsilon_j - \alpha_j \ \Delta T) \ . \tag{7.56}
$$

Equation (7.56) shows that if ε_i is equal to ε_i^t the material is free of stresses. Then the entire deformations are due to thermal expansion.

So far the description is three-dimensional and concerns generally anisotropic materials. Further on the theory will be limited to orthotropic laminae in plane stress. With respect to the axes of symmetry it follows that:

$$\alpha_4 = \alpha_5 = \alpha_6 = 0 \ . \tag{7.57}$$

The thermal expansion in x_3-direction is assumed to be irrelevant. Analogous to equation (6.33) the in-plane stress-strain relations with respect to the axes of symmetry are:

$$\sigma_i = \sum_j Q_{ij} \left(\varepsilon_j - \alpha_j \, \Delta T \right) ; \quad i, j \in \{1,2,6\} \ . \tag{7.58}$$

Consider equation (7.58) applied to the kth lamina of a laminate. With respect to the principal directions of the laminate these relations are:

$$\begin{bmatrix} \sigma_1' \\ \sigma_2' \\ \sigma_6' \end{bmatrix}_k = \begin{bmatrix} Q_{11}' & Q_{12}' & Q_{16}' \\ & Q_{22}' & Q_{26}' \\ \text{symm} & & Q_{66}' \end{bmatrix}_k \begin{bmatrix} \varepsilon_1' - \alpha_1' \, \Delta T \\ \varepsilon_2' - \alpha_2' \, \Delta T \\ \varepsilon_6' - \alpha_6' \, \Delta T \end{bmatrix}_k , \tag{7.59a}$$

where α_1', α_2' and α_6' are the thermal expansion components of the orthotropic lamina with respect to the principal directions of the laminate.

The right term of equation (7.59a) can be split up:

$$\begin{bmatrix} \sigma_1' \\ \sigma_2' \\ \sigma_6' \end{bmatrix}_k = \begin{bmatrix} Q_{11}' & Q_{12}' & Q_{16}' \\ & Q_{22}' & Q_{26}' \\ & & Q_{66}' \end{bmatrix}_k \begin{bmatrix} \varepsilon_1' \\ \varepsilon_2' \\ \varepsilon_6' \end{bmatrix}_k - \begin{bmatrix} Q_{11}' & Q_{12}' & Q_{16}' \\ & Q_{22}' & Q_{26}' \\ & & Q_{66}' \end{bmatrix}_k \begin{bmatrix} \alpha_1' \\ \alpha_2' \\ \alpha_6' \end{bmatrix}_k \Delta T. \tag{7.59b}$$

Substitution in equation (7.6) results in the following expressions:

$$\begin{bmatrix} N_x \\ N_y \\ N_{xy} \end{bmatrix} = \begin{bmatrix} A_{11} & A_{12} & A_{16} \\ & A_{22} & A_{26} \\ & & A_{66} \end{bmatrix} \begin{bmatrix} \varepsilon_x^0 \\ \varepsilon_y^0 \\ \varepsilon_{xy}^0 \end{bmatrix} + \begin{bmatrix} B_{11} & B_{12} & B_{16} \\ & B_{22} & B_{26} \\ & & B_{66} \end{bmatrix} \begin{bmatrix} \kappa_x \\ \kappa_y \\ \kappa_{xy} \end{bmatrix} - \begin{bmatrix} N_x^t \\ N_y^t \\ N_{xy}^t \end{bmatrix} , \tag{7.60}$$

with

$$
\begin{bmatrix} N_x^t \\ N_y^t \\ N_{xy}^t \end{bmatrix} = \sum_{k=1}^{n} \int_{z_{k-1}}^{z_k} \begin{bmatrix} Q_{11}' & Q_{12}' & Q_{16}' \\ & Q_{22}' & Q_{26}' \\ & & Q_{66}' \end{bmatrix}_k \begin{bmatrix} \alpha_1' \\ \alpha_2' \\ \alpha_6' \end{bmatrix}_k \Delta T \, dz .
\tag{7.61}
$$

The so-called thermal forces per unit length, $\{N^t\}$, are not real forces. But to ensure that the laminate is not deformed, i.e. $\{\varepsilon^0\} = \{0\}$ and $\{\kappa\} = \{0\}$, external forces per unit length, $\{N\}$, equal to $\{N^t\}$ and acting in the opposite direction are needed.

An example to illustrate this theoretical description now follows. The thermal expansion components of a transversely isotropic unidirectional plate are $\alpha_L > 0$, $\alpha_T \gg 0$, and $\alpha_{LT} = 0$. The temperature of the plate is increased by ΔT. Equation (7.61) gives the longitudinal thermal force per unit length as:

$$
N_L^t = (Q_{11} \, \alpha_L + Q_{12} \, \alpha_T) \, \Delta T \cdot h ,
$$

where h is the plate thickness. An external force $N_L = -N_L^t$ is needed to annul the longitudinal thermal expansion.

Substitution of equation (7.59b) into equation (7.10) leads to equations for the moments per unit length of the edge:

$$
\begin{bmatrix} M_x \\ M_y \\ M_{xy} \end{bmatrix} = \begin{bmatrix} B_{11} & B_{12} & B_{16} \\ & B_{22} & B_{26} \\ & & B_{66} \end{bmatrix} \begin{bmatrix} \varepsilon_x^0 \\ \varepsilon_y^0 \\ \varepsilon_{xy}^0 \end{bmatrix} + \begin{bmatrix} D_{11} & D_{12} & D_{16} \\ & D_{22} & D_{26} \\ & & D_{66} \end{bmatrix} \begin{bmatrix} \kappa_x \\ \kappa_y \\ \kappa_{xy} \end{bmatrix} - \begin{bmatrix} M_x^t \\ M_y^t \\ M_{xy}^t \end{bmatrix} ,
\tag{7.62}
$$

with

$$
\begin{bmatrix} M_x^t \\ M_y^t \\ M_{xy}^t \end{bmatrix} = \sum_{k=1}^{n} \int_{z_{k-1}}^{z_k} \begin{bmatrix} Q_{11}' & Q_{12}' & Q_{16}' \\ & Q_{22}' & Q_{26}' \\ & & Q_{66}' \end{bmatrix}_k \begin{bmatrix} \alpha_1' \\ \alpha_2' \\ \alpha_6' \end{bmatrix}_k \Delta T \, dz .
\tag{7.63}
$$

A similar remark as made in relation to the thermal forces can be made in relation to these thermal moments.

The thermal forces and moments can be added to the mechanical forces and moments:

$$\begin{bmatrix} N_x^f \\ N_y^f \\ N_{xy}^f \end{bmatrix} = \begin{bmatrix} N_x + N_x^t \\ N_y + N_y^t \\ N_{xy} + N_{xy}^t \end{bmatrix} = [A] \{\varepsilon^0\} + [B] \{\kappa\} . \qquad (7.64)$$

$$\begin{bmatrix} M_x^f \\ M_y^f \\ M_{xy}^f \end{bmatrix} = \begin{bmatrix} M_x + M_x^t \\ M_y + M_y^t \\ M_{xy} + M_{xy}^t \end{bmatrix} = [B] \{\varepsilon^0\} + [D] \{\kappa\} . \qquad (7.65)$$

The fictitious forces and moments per unit length, $\{N^f\}$ and $\{M^f\}$, respectively, can be handled in the laminate theory as if being the resultants of only mechanical loads.

The equations (7.64) and (7.65) can be inverted. The result is:

$$\begin{bmatrix} \{\varepsilon^0\} \\ ... \\ \{\kappa\} \end{bmatrix} = \begin{bmatrix} [a] & : & [b] \\ .... & : & .... \\ [b^T] & : & [d] \end{bmatrix} \begin{bmatrix} \{N^f\} \\ .... \\ \{M^f\} \end{bmatrix} . \qquad (7.66)$$

If the mechanical load and the thermal load, i.e. ΔT, are given, equation (7.66) will yield the deformations of the midplane of the laminate, and with that the deformations of the laminae with the help of equation (7.4). The thermal deformations of the lamina following on $\{\alpha'\}_k$ and ΔT should be subtracted. Substitution of the resulting mechanical deformations into equation (7.3) gives the real stresses in the kth lamina with respect to the principal directions of the laminate.

The swelling due to moisture absorption can be handled in an analogous manner. A difference in practice is the fact that an equilibrium in temperature is reached much faster than an equilibrium in moisture content, both internally and in contact with the environment.

Both phenomena are an important part concerning the behaviour of FRP. After curing at an elevated temperature stresses will occur - due to shrinkage - during cooling to room temperature. And after that during every change in temperature. Because the thermal expansion is anisotropic, it is necessary to construct laminates that are strictly symmetric in order to avoid curving and twisting of the laminate simply due to cooling to room temperature.

7.7 INTERLAMINAR STRESSES AND FREE EDGE EFFECTS

In some applications of a FRP laminate - e.g. a leaf spring - a shear force normal to the laminate is present, sometimes in a prominent way. Such a shear force introduces shear stresses in a cross-section perpendicular to the midplane. Then on equilibrium grounds, equal shear stresses act in planes parallel to the midplane (and perpendicular to the cross-section). Some of those planes are interfaces. The so-called interlaminar stresses operating in the interfaces can play a very important part with regard to the strength of the laminate, especially because the strength of an interface is much smaller than the in-plane strength of the laminae. However, the classical laminate theory is based on the assumption that only in-plane stresses are acting in the laminae. That means the existence of interlaminar stresses is neglected. In such cases the CLT is not sufficient. It will be necessary to extend the CLT to a three-dimensional theory.

Another way to tackle the problem is to make a finite element analysis (FEA) using 3-D orthotropic elements. For a stiffness analysis, a knowledge of the laminate stiffnesses perpendicular to the laminate is necessary. A strength analysis is more complicated. The failure criteria usually formulated at lamina level with respect to in-plane stress have to be extended to 3-D at laminate level and to include failure criteria for the interfaces.

The CLT is also based on the assumption that a normal line on the midplane stays perpendicular to that plane in the deformed state. The laminae have to deform collectively. For this forced interaction force transfer between the laminae near the edges is required. These interlaminar stresses can be excessive and introduce delamination. The presence of these stresses will be shown by means of some qualitative examples.

A simple cross-ply laminate is constructed symmetrically of four identical unidirectional laminae: $[0/90]_s$. Out of this laminate a strip is sawn with the length direction (x-axis) corresponding to one of the fibre directions (0°-direction). The strip is pulled in x-direction. Because the longitudinal modulus of elasticity, E_L, is much bigger than the transverse one, E_T, according to the reciprocal equation (6.25) the major Poisson's ratio, v_{LT}, is much bigger than the minor one, v_{TL}. Both 0° laminae and 90° laminae are equally strained in x-direction. Under the influence of this elongation all the laminae will contract in y-direction, see figure 7.10, but the 0° laminae will contract more than the 90° laminae ($v_{LT} > v_{TL}$). The physical explanation is that the stiff fibres in y-direction of the 90° laminae are not willing to contract in length direction. But due to being glued together the laminae are forced to contract by the same amount. The consequences are a tensile stress in y-direction in the 0° laminae, which decreases the contraction, and a compressive stress into the 90° laminae, which increases the contraction. These stresses can be calculated by the CLT. However, it is impossible for these stresses to act up to the free edges of the strip, because the free edges are without stresses. Equilibrium in y-direction of one half of the 4th lamina requires the presence of interlaminar shear stress σ_{zy}, see figure 7.10. Near the edge σ_{zy} must decrease to zero. To make equilibrium of moments within the yz-plane of that half of the 4th lamina a normal stress σ_z will act on the interface between the 3rd and the 4th lamina. This interlaminar normal stress is positive and maximal at the free edge. Going into the laminate σ_z decreases and changes into compression. Further in, the compression decreases again to zero. Interlaminar tension and compression must be in equilibrium with each other. Between the 2nd and the 3rd lamina the normal stress is even stronger.

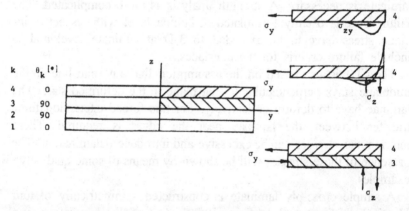

Figure 7.10. Cross-section (yz-plane) of a strip out of a cross-ply laminate [0/90]$_s$ pulled in x-direction (0°). Interlaminar stresses σ_{zy} and σ_z realise equilibrium. At the free edge the interlaminar normal stress σ_z is a tensile stress.

An angle-ply laminate of four identical unidirectional laminae is symmetrically constructed: $[\theta/-\theta]_s$. The length direction of a strip (x-axis) corresponds to the 0° direction of the laminate. The strip is pulled in length direction. Under the influence of the elongation both kinds of laminae will contract in the same measure, because the

Poisson's ratios v_{xy} are equal. In this respect the laminae are matched. But due to difference in sign of the coefficients of mutual influence,

$$\rho'_{16}(-\theta) = -\rho'_{16}(\theta) \, ,$$

the 3rd and the 4th lamina will shear in opposite directions, as equation (6.62) makes clear, see figure 7.11. These effects neutralise

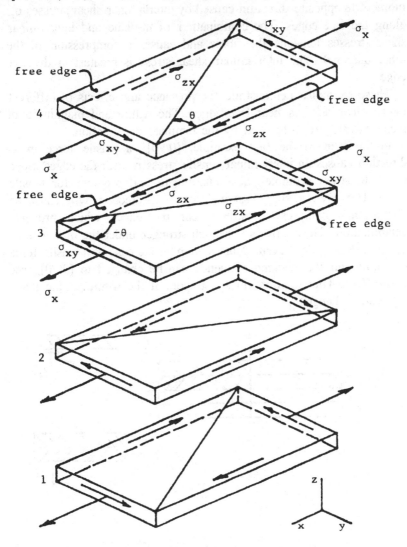

Figure 7.11. Exploded view of a part of a strip out of an angle-ply laminate $[\theta/-\theta]_s$ pulled in x-direction (0°). In the neighbourhood of the free edges interlaminar shearing stresses σ_{zx} along the edges are caused.

each other. The result is that there is no shearing between x and y-axis of the laminate. But to realise that an in-plane shear strain σ_{xy} will act in the laminae, in both kinds of laminae in opposite direction, see figure 7.11. These shear stresses can be calculated by CLT. Near the free edges σ_{xy} must decrease to zero. The resulting moment of the shear stresses σ_{xy} in the 4th lamina must be compensated by a moment in opposite direction caused by interlaminar shear stresses σ_{zx} along the free edges. The combination of in-plane and interlaminar shear stresses tensions the fibres and causes a compression of the other diagonal. The interlaminar shear stress is greatest at the free edge.

These examples demonstrate the presence and effects of different interlaminar stresses near free edges. The nature and magnitude of these stresses are influenced by the laminate construction.

In the strip of the first example, $[0/90]_s$, a tensile force in x-direction causes an interlaminar tensile stress σ_z near the edges. Such a tensile stress promotes delamination (a real problem for tensile tests). However, a compressive force in x-direction will cause an interlaminar compressive stress near the edges which prevents delamination. These effects are much stronger under the influence of dynamic loads. To prevent delamination in the case of a tensile force in x-direction the stacking sequence can be changed to $[90/0]_s$, see figure 7.12. This solution does not work if the nature of the loads repeatedly changes.

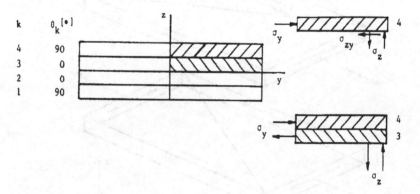

Figure 7.12. Cross-section (yz-plane) of a strip out of a cross-ply laminate $[90/0]_s$ pulled in x-direction (0°). At the edges the interlaminar normal stress σ_z is compressive.

In the second example the magnitude of the interlaminar stress depends on the amount by which the strain ratios ρ_{16}' of successive laminae differ from each other. Figure 7.13 shows the values of ρ_{16}' of unidirectional glass/polyester as a function of the orientation angle. The extreme values are near $\theta \approx \pm 15°$. For that reason the mismatch with respect to the coefficients of mutual influence is maximal for $[15/-15]_s$. Laminae oriented in $0°$-direction and in $\pm 60°$-direction match in respect to ρ_{16}' and also to the Poisson's ratios v_{12}', see figure 6.11. Maybe a smart combination of laminae with those orientation angles brings the solution to the problem. (See also [2]).

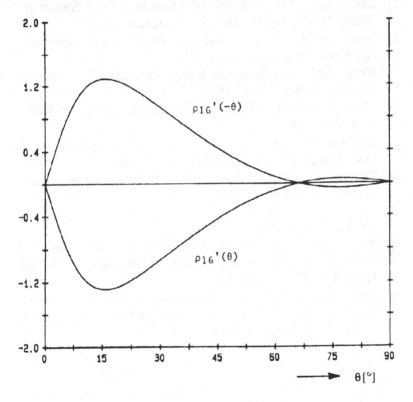

Figure 7.13. Mismatch of strain ratios ρ_{16}' of two laminae with orientation θ and -θ, respectively (unidirectional glass/polyester, $V_f \approx 0.6$). The mismatch is extreme at $\theta \approx 15°$.

If a difference between anisotropic expansion coefficients of successive laminae exists, similar effects as mentioned above will occur, either due to change of temperature or due to an increase or decrease of moisture content. In hybrid laminates reinforced with layers of fibres with very different thermal expansion coefficients those effects happen in an intensified measure.

Calculations, e.g., by FEA, have proved that interlaminar stresses near the free edges only act in a small zone with a width of about the laminate thickness. It pays to avoid free edges and to make integral constructions. To avoid delamination edges can be reinforced, promoting interaction of forces between the laminae near these edges.

7.8 BIBLIOGRAPHY

1] Lekhnitskii, S.G., "Theory of Elasticity of an Anisotropic Elastic Body", Holden Day, San Francisco, 1963.

2] Sendeckyj, G.P. (ed.), "Mechanics of Composite Materials", Academic Press, New York, 1974.

3] Jones, R.M., "Mechanics of Composite Materials", McGraw-Hill Kogakusha, Tokyo, 1975.

4] Christensen, R.M., "Mechanics of Composite Materials", Wiley, New York, 1979.

5] Agarwal, B.V., Broutman, L.J., "Analysis and Performance of Fiber Composites", Wiley, New York, 1980.

6] Tsai, S.W., Hahn, H.T., "Introduction to Composite Materials", Technomic, Westport, 1980.

7.9 REFERENCES

1] Bartholomew, P., "Ply Stacking Sequences for Laminated Plates having In-plane and Bending Orthotropy", Fibre Sci. Tech., 10, 1977. pp 239-253.

2] Herakovich, C.T., "On the Relationship between Engineering Properties and Delamination of Composite Materials", J. Comp. Mater., 15, 1981. pp 336-348.

8 THEORY OF SANDWICH BEAMS AND PLATES

8.1 INTRODUCTION

8.1.1 What is a 'Structural Sandwich'?

The structural steel I-beam (universal beam in the UK) has properties that are familiar to most engineers. Most of the material (perhaps 70-80%) is concentrated in the flanges, where it is most effective in contributing to the bending stiffness and bending strength of the beam. The consequence of this concentration is that the shear stress in the web is relatively high and it does not vary much across the depth of the web; see figure 8.1. Where there is shear stress there is

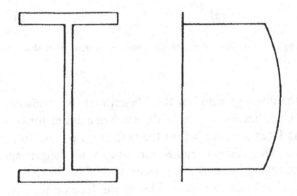

Figure 8.1.   a) Steel I-beam,   b) Shear stress in web.

also shear strain. The shear strain leads to additional deflection under load. This is usually small in comparison with the ordinary bending deflection. Consequently ordinary steel I-beams can be analysed quite effectively by using conventional bending theory ('plane sections remain plane and perpendicular to the longitudinal axis').

A structural sandwich beam, figure 8.2, consists of two thin, strong, stiff faces attached (usually by adhesive) to opposite sides of a thick low density material (the core). The faces take the place of the flanges

in the I-beam; as in the I-beam they provide most of the bending
stiffness and bending strength. The core takes the place of the web.
It is possible to imagine that the material in the web of the I-beam is
expanded by means of air-bubbles to fill the space between the
flanges; see figure 8.2.

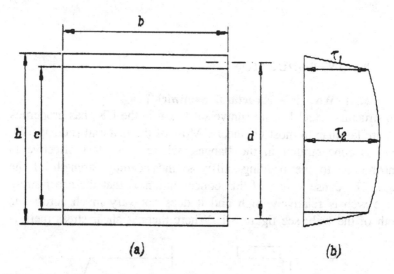

(a)   (b)

Figure 8.2. a) Cross-section of sandwich beam, b) Variation of shear stress across
depth of beam.

So far the analogy between the I-beam and the sandwich beam is
exact, but complications arise in the sandwich beam for two reasons.

First, the faces are attached to the core at all points, so the normal
limits on width/thickness ratios for flanges no longer apply. It is
therefore tempting to make the faces very thin, in the interests of
efficiency. This leads to new problems: the face on the compression
side may suffer a new type of short-wavelength instability in some
circumstances (wrinkling instability, figure 8.3a); and transverse loads
may cause local deformation of the face as the underlying core
subsides (indentation failure, figure 8.3b). Both of these kinds of
failure occur when the core is excessively flexible across the thickness
of the sandwich, thereby failing to keep the faces the correct fixed
distance apart.

The second complication that occurs in sandwich beams (but not in
I-beams) is that the core and the faces can be made from different
materials. This encourages designers to use the lowest possible density
for the core material. This means that the shear modulus (modulus of

rigidity) of the core is reduced and the shear strains in it become larger. It is not always possible to ignore the effects of these shear strains on the deflections of beams and plates, and on the critical loads of columns and panels.

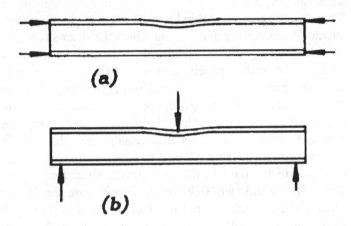

Figure 8.3. a) Short wavelength wrinkling, b) Indentation failure.

An efficient sandwich will have thin faces and a core which is not too flexible in shear; this is generally the case in aerospace structures. In other applications, practical considerations sometimes dictate the choice of materials; cases arise in which the faces are somewhat thick while the core is not especially stiff. Analysis of such sandwich beams is somewhat complex; solutions have been available in the literature for over twenty years, but they are not well known to designers. Some guidance is given below.

8.2 BENDING OF BEAMS

8.2.1 Classification of Sandwich Panels
Perhaps the most useful piece of information for a designer is not a collection of differential equations, but a clear understanding of the three different regimes of structural sandwich behaviour, and of the kind of analysis appropriate to each one. The three regimes are defined below.

(a) **Composite Beam Behaviour**: In this situation the core is relatively stiff, the core shear strains are small, and the effects of shear deformation can be ignored. The sandwich can be treated in exactly the same way as a conventional beam, except that section properties need to be calculated as for any other sort of composite beam. For example, the core can be replaced by an equivalent width of face material when calculating cross-sectional properties.

(b) **Thin-face Sandwich Behaviour**: This is the common case of structural sandwich behaviour, in which the shear strains in the core lead to significant additional deflections in beams and transversely loaded panels, and to reductions in the critical loads in columns and in panels with edge-loading. The direct stresses in the faces, arising from bending actions, are generally not much affected by these additional deflections. In a sandwich of this type it is also usually true that the faces contribute most of the resistance to bending moment while the core contributes most of the resistance to shear force.

(c) **Thick-face Sandwich Behaviour**: In this case the faces are relatively thick and the core is relatively flexible. The applied load is resisted partly by behaviour analogous with that of a thin-face sandwich; and partly by the local bending of the faces themselves. In fact the easiest way to handle the problem is to imagine the load to be split into two parts, one resisted by the sandwich-action, the other by the local bending of the faces. Thick-face sandwich behaviour is also applicable to sandwiches with corrugated faces; see figure 8.4. It can be shown that the equations for thick-face sandwich behaviour are also applicable to many other types of structure, such as parallel-chord trusses and shear-wall buildings.

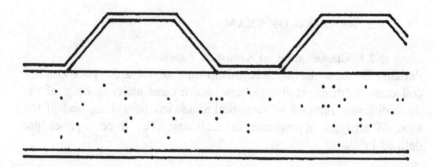

Figure 8.4. Sandwich with corrugated face.

Methods of analysis for all three kinds of behaviour will be described briefly below. It is most important to define the boundaries between them, so that designers will know when a particular kind of behaviour is likely to occur.

8.2.2 Composite Beams

Methods for calculating the section properties of composite beams, using conventional beam theory, are well-known and will not be repeated here. But there are some interesting facts arising from the theory worth mentioning. The notation used in the following is as in figure 8.2.

First, how much does the shear stress vary across the depth of the core? If τ_1 is the shear stress at the interface and τ_2 is the shear stress at the middle plane of the sandwich, then the ratio

$$\frac{(\tau_2 - \tau_1)}{\tau_1} \text{ is approximately } \frac{1}{4}\frac{c}{t}\frac{E_c}{E_f} \qquad (8.1)$$

where E_c, E_f are the Young's moduli of the core and faces respectively.

Secondly, how much does the core contribute to the flexural rigidity of the sandwich? If D_1 is the contribution of the core and D_2 is the contribution of the faces, then the ratio

$$\frac{D_1}{D_2} \text{ is approximately } \frac{1}{6}\frac{c}{t}\frac{E_c}{E_f}. \qquad (8.2)$$

These two expressions are very similar. It can be said that when one expression is negligible the other one is negligible also. This demonstrates the important conclusion that:

When the core makes a negligible contribution to the flexural stiffness of the sandwich, the shear stress is approximately uniform through the depth of the core.

It might also be asked at what point it becomes unwise to treat a sandwich as a simple composite beam. In the case of a simply-supported sandwich beam with a central point load, it can be shown that the ratio of shear deflection, Δ_s, to bending deflection, Δ_b, is approximately

$$\frac{\Delta_s}{\Delta_b} = 6\frac{t}{c}(\frac{c}{L})^2 \frac{E_f}{G_c} \tag{8.3}$$

where L is the beam's span and G_c is the shear modulus of the core.

This gives some guidance as to when shear deflections are likely to become important. The expression shows that a low shear modulus, or a deep short beam, or a sandwich with a relatively thick face, may all tend to produce significant shear deflections.

The same expression can be written more compactly (and more precisely) as

$$\frac{\Delta_s}{\Delta_b} = 12 \frac{D}{L^2 D_Q} \tag{8.4}$$

where D is the overall flexural rigidity of the sandwich and D_Q is the shear stiffness (roughly the cross-sectional area of the core multiplied by the shear modulus of the core; the exact value is given later). Expressions for D and D_Q are given later, in section 8.2.3, equation (8.6) and in section 8.2.5, equation (8.31) respectively.

It will be seen that the quantity

$$\frac{D}{L^2 D_Q} \tag{8.5}$$

occurs repeatedly in sandwich formulae. It is one of the fundamental sandwich parameters defining the ratio of the flexural stiffness to the shear stiffness, made non-dimensional by the inclusion of the (span)^2 term.

8.2.3 Thin-Face Sandwich Beams

It is convenient to begin with the case of equal faces. Using the notation in figure 8.2, the flexural stiffness due to the faces is

$$D = E_f \frac{btd^2}{2} + D_f \tag{8.6}$$

where $D_f = \dfrac{E_f b t^3}{6}$.

D_f, which arises from the second moments of area of the faces about their own centroids, can usually be neglected. If the beam is a wide one, bending will be cylindrical and the faces will not be able to expand or contract laterally; in that case, for isotropic materials, Young's modulus should be replaced by

$$\frac{E_f}{(1-v^2)} \tag{8.7}$$

where v is Poisson's ratio.

Bending deflections and stresses can be calculated in the usual way; if the faces are thin it is usually sufficiently accurate to calculate the direct stresses at the middle planes of the faces.

The additional deflections due to shear strains in the core are easily calculated. Consider, for example, the simply-supported beam in figure 8.5. It carries a central point load W and the shear force

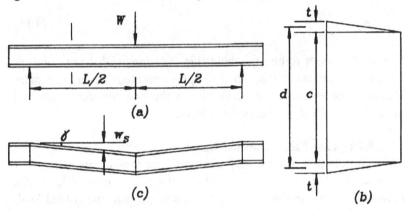

Figure 8.5. a) Simply supported beam with central point load, b) Effective shear stress in core, c) Shear deformation of beam.

everywhere is $Q = W/2$. The inset diagram shows the distribution of shear stress from top to bottom of the beam. Since this is a thin-face sandwich, the shear stress in the core is effectively uniform (compare figure 8.2b); it can also be shown to decrease uniformly across the thickness of each face to zero at the free surface. If the shear stress in the core is τ, the shear force, Q, is given by

$$Q = bc\tau + 2\left(\frac{bt\tau}{2}\right) = b(c + t)\tau = bd\tau. \tag{8.8}$$

The shear strain (γ) in the core is

$$\gamma = \frac{\tau}{G_c}. \tag{8.9}$$

This shear strain is also the gradient of the shear displacement, as shown in figure 8.5c. It is easy to show in this example that

$$\Delta_s = \gamma \frac{L}{2} = \frac{Q}{bdG_c}\frac{L}{2} = \frac{WL}{4A_cG_c} \tag{8.10}$$

where $A_c = bd$ is the effective area of the core.

It is worth noting that the shear deflection is the integral of the shear strain, which is proportional to the shear force. Since the bending moment is also the integral of the shear force, it is not surprising that the expression for the shear deflection is the same as the expression for the bending moment, divided by A_cG_c. Thus, the shear deflection at the centre of a simply-supported beam with a uniformly distributed load with a total magnitude W is:

$$\Delta_s = \frac{WL}{8A_cG_c}. \tag{8.11}$$

Little else needs to be said about thin-face sandwich beams, except perhaps to emphasise that, in a simply-supported beam, the shear deflection occurs quite independently of the bending deflection and has no major effect on the stresses in the faces.

8.2.4 Thin-Face Sandwich Columns
Sandwich columns are not very common, but it is worth noting that the shear flexibility of the core reduces the elastic critical load of a sandwich column. In the case of a pin-ended column, the critical load, P_{cr} becomes:

$$P_{cr} = \frac{P_E}{1 + \dfrac{P_E}{A_cG_c}}. \tag{8.12}$$

Where $P_E = \dfrac{\pi^2 D}{L^2}$ is the Euler load. $\tag{8.13}$

The expression for the critical load can also be written as an interaction formula:

$$\frac{1}{P_{cr}} = \frac{1}{P_E} + \frac{1}{A_cG_c}. \tag{8.14}$$

In either formula it is easy to see that, as the shear stiffness term A_cG_c diminishes, the critical load falls short of the usual Euler value. In many texts the authors consider what happens as the core shear modulus G_c becomes very small. The second term on the right-hand side of equation (8.14) becomes dominant, so the first term can be neglected, leading to the result:

$$P_{cr} = A_c G_c. \qquad (8.15)$$

On the face of it, this corresponds to a kind of shear instability as in figure 8.6. Examples of this kind of instability (shear buckling) have been reported many times. Unfortunately equation (8.15) is not much help in adding to our understanding of sandwich behaviour in extreme circumstances because it contains a logical contradiction; the presence of shear stress in the core implies the presence of faces, but the faces have become vanishingly thin. This contradiction is resolved when thick-face sandwich theory is used, as described below.

Figure 8.6. Shear buckling (Diagrammatic).

8.2.5 Thick-Face Sandwich Beams
In order to understand the theory of thick-face sandwich beams, it is essential to discard the simplistic notion that

'total deflection = bending deflection plus shear deflection'.

In order to avoid confusion we will refer to primary displacements and secondary displacements, which are defined rather differently.

Figure 8.7a shows a short length of a sandwich beam in the undeflected state. On the side of the beam is marked a plane cross-section *abcdefg*. It is assumed that when load is applied, the point *d*, which is at the half-height of the beam, moves vertically downwards (symmetry requires this). There are two ways in which this movement may be accomplished.

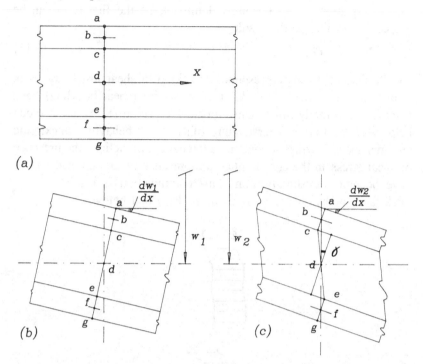

(a)

(b)

(c)

Figure 8.7. a) Short length of sandwich beam, b) Primary displacement,
c) Secondary displacement.

One way is shown in figure 8.7b. Here the beam bends as a whole,
the point d descends vertically, and the cross-section $abcdefg$ rotates
while still remaining perpendicular to the longitudinal axis of the
beam. This will be called the primary displacement (w_1); evidently it
conforms to all the usual rules of the 'Engineer's Bending Formula'.
But not all of the applied load is resisted by the beam in this manner.

Figure 8.7c shows an alternative mode of deformation, in which the
points b and f, which lie at the centroids of the faces, move vertically
downwards. In effect, the faces bend about their own centroids but do
not shorten or stretch. Since there is no overall movement of the faces
along the length of the beam (to left or right in the diagram), the
mean direct stress in each face is zero. This type of deformation will
be called the secondary displacement (w_2). The rotation of the faces
(associated with their bending) causes a shear strain (γ) in the core as
shown. From the geometry of the diagram, the shear strain (γ) is
related to the slope or gradient of the beam by the equation

$$\gamma = \frac{d}{c}\frac{dw_2}{dx}. \qquad (8.16)$$

The primary and secondary displacements are independent of each other.

The problem to be solved is this. Both the primary and secondary displacement patterns provide bending moments and shear forces capable of resisting a part of the applied load; in what proportion do they contribute to the overall bending moments and shear forces?

There are at least two ways of solving the problem. In a 1969 book on sandwich panels [1], compatibility and equilibrium considerations were used to show how the basic differential equations of the problem can be formulated. An alternative, and more general, formulation of the differential equations, based on energy considerations, was described in an internal report of 1971 [2].

The differential equations can be formulated in various ways. One convenient form, in which the unknowns are the primary and secondary displacements w_1 and w_2 (and their derivatives) is:

$$D_f(D-D_f)w_1^{vi} + \{P(D-D_f)-DD_Q\}w_1^{iv} - PD_Q w_1^{ii} \qquad (8.17)$$
$$= -qD_Q$$

$$D_f(D-D_f)w_2^{vi} + \{P(D-D_f)-DD_Q\}w_2^{iv} - PD_Q w_2^{ii} \qquad (8.18)$$
$$= +q^{ii}(D-D_f).$$

In equations (8.17) and (8.18) a superscript denotes the order of the derivative, for example $w_1^{iv} = d^4w_1/dx^4$; q is the local transverse load on the beam and P is an end load applied along the beam.

The solution of these equations for particular cases is tedious, involving numerous boundary conditions, together with a number of hyperbolic functions. Solutions for simply supported beams with (a) central point load and (b) uniformly distributed load are given in [1,2] and [3], but even the solutions are awkward to apply. To simplify the calculations, a set of tabulated coefficients (S) can be used, as explained below for two common cases.

In each of these cases the beam is simply supported, with span L, and an overhang at each end, and it supports a total load W. In the first case the load W is concentrated at the mid-point. In the second case the load is distributed uniformly over the span L. The quantities of greatest interest to the designer are the maximum deflection, the maximum bending moment and shear force, and the associated stresses. Before these can be obtained it is necessary to show how the coefficients (S) appear in the relevant formulae.

A) Central point load, W

Total central deflection (primary and secondary combined):

$$w = \frac{WL^3}{48D} + \frac{WL}{4D_Q}\left(1 - \frac{D_f}{D}\right)^2 S_1.$$ (8.19)

Primary and secondary deflections separately are:

$$w_1 = \frac{WL^3}{48D}S_1'$$ (8.20)

$$w_2 = \frac{WL^3}{48D_f}(1 - S_1').$$ (8.21)

Primary shear force Q_1, which occurs between the support and the mid-point (regardless of sign):

$$(Q_1)_{max} = \frac{W}{2}S_2.$$ (8.22)

The primary and secondary bending moments at midspan are:

$$M_1 = \frac{WL}{4}S_3$$ (8.23)

$$M_2 = \frac{WL}{4}(1 - S_3).$$ (8.24)

B) Uniformly distributed load, total value W

Total central deflection (primary and secondary combined):

$$w = \frac{5}{384}\frac{WL^3}{D} + \frac{WL}{8D_Q}\left(1 - \frac{D_f}{D}\right)^2 S_4.$$ (8.25)

Primary and secondary deflections separately are:

$$w_1 = \frac{5}{384} \frac{WL^3}{D} S_4'$$
(8.26)

$$w_2 = \frac{5}{384} \frac{WL^3}{D_f} (1 - S_4').$$
(8.27)

Primary shear force Q_1, which occurs between the support and the mid-point (regardless of sign):

$$(Q_1)_{max} = \frac{W}{2} S_5.$$
(8.28)

The primary and secondary bending moments at midspan are:

$$M_1 = \frac{WL}{8} S_6$$
(8.29)

$$M_2 = \frac{WL}{8} (1 - S_6).$$
(8.30)

C) Calculation of the coefficients (S)

A computer program is available with the author which allows the coefficients S, the deflections, bending moments, shear forces and stresses to be obtained for any individual case.

It is also possible (but less convenient) to solve any particular problem by using the S-coefficients provided in table 8.1. The solution proceeds in the following steps.

i)   Write down the cross-sectional dimensions of the beam (as in figure 8.2) and note the values of E_f and G_c. If G_c is not available from the manufacturer of the core material, and it cannot be found in the literature, it will need to be obtained from tests on samples.

ii) Calculate the overall flexural rigidity D of the complete sandwich, treating it as a simple beam. For sandwiches with equal faces, equation (8.6) can be used. For sandwiches with unequal faces, the value of D can be worked out from first principles in the usual way.

iii) Calculate the local flexural stiffnesses of the two faces and add them together to obtain D_f. This is the second term in equation (8.6).

iv) Calculate the shear stiffness of the core, D_Q where

$$D_Q = \frac{b\,d^2}{c}\,G_c. \tag{8.31}$$

(Notice that the effective area of the core, mentioned previously as $A_c = bdG_c$, is an approximation to D_Q).

Table 8.1. Values of S_1-S_6 for very large overhang L_1 ($\phi = 10$)

θ	S_1	S_1'	S_2	S_3	S_4	S_4'	S_5	S_6
0.1	0.0031	0.0716	0.0045	0.0484	0.0038	0.0762	0.0030	0.0642
0.2	0.0115	0.1369	0.0106	0.0937	0.0142	0.1454	0.0110	0.1238
0.3	0.0241	0.1966	0.0342	0.1361	0.0297	0.2082	0.0227	0.1792
0.4	0.0399	0.2511	0.0559	0.1758	0.0490	0.2654	0.0370	0.2306
0.5	0.0582	0.3011	0.0807	0.2131	0.0711	0.3175	0.0533	0.2784
0.6	0.0784	0.3469	0.1076	0.2480	0.0952	0.3651	0.0709	0.3228
0.7	0.0998	0.3889	0.1360	0.2808	0.1208	0.4085	0.0893	0.3641
0.8	0.1221	0.4275	0.1653	0.3117	0.1471	0.4482	0.1083	0.4025
0.9	0.1450	0.4630	0.1951	0.3406	0.1739	0.4846	0.1275	0.4382
1.0	0.1681	0.4957	0.2251	0.3679	0.2008	0.5180	0.1468	0.4715
1.5	0.2809	0.6255	0.3703	0.4821	0.3294	0.6486	0.2387	0.6070
2.0	0.3808	0.7144	0.4977	0.5677	0.4393	0.7364	0.3183	0.7030
2.5	0.4643	0.7771	0.6032	0.6328	0.5282	0.7972	0.3845	0.7719
3.0	0.5328	0.8224	0.6884	0.6833	0.5987	0.8403	0.4389	0.8220
3.5	0.5886	0.8559	0.7561	0.7229	0.6545	0.8718	0.4839	0.8589
4.0	0.6341	0.8811	0.8095	0.7546	0.6988	0.8952	0.5215	0.8864
4.5	0.6716	0.9005	0.8514	0.7802	0.7344	0.9130	0.5532	0.9073
5.0	0.7027	0.9157	0.8841	0.8013	0.7632	0.9267	0.5804	0.9232
6.0	0.7508	0.9374	0.9296	0.8337	0.8065	0.9462	0.6246	0.9454
7.0	0.7860	0.9519	0.9573	0.8573	0.8370	0.9590	0.6591	0.9595
8.0	0.8126	0.9619	0.9741	0.8750	0.8595	0.9678	0.6870	0.9688
9.0	0.8334	0.9691	0.9843	0.8889	0.8766	0.9740	0.7101	0.9753
10.0	0.8500	0.9745	0.9905	0.9000	0.8900	0.9786	0.7295	0.9800
15.0	0.9000	0.9880	0.9992	0.9333	0.9289	0.9901	0.7947	0.9911
20.0	0.9250	0.9931	0.9999	0.9500	0.9475	0.9943	0.8324	0.9950
30.0	1.0000	0.9967	1.0000	0.9667	0.9656	0.9974	0.8753	0.9978
40.0	1.0000	0.9981	1.0000	0.9750	0.9744	0.9985	0.8995	0.9988

v) Calculate the non-dimensional sandwich parameters $D/D_Q L^2$ and D_f/D.

vi) Calculate the value of θ from the following formula:

$$\theta = 0.5 \left[\left(\frac{D}{L^2 D_Q} \right) \frac{D_f}{D} \left(1 - \frac{D_f}{D} \right) \right]^{-\frac{1}{2}}. \qquad (8.32)$$

vii) Use this to find the values of the relevant coefficients (S) in table 8.1.

viii) Calculate the relevant deflections, bending moments and shear forces from equations (8.19-8.30) as required.

The bending and shear stresses can be calculated as in the example below.

D) Practical considerations

Table 8.1 is based on the assumption that there is a substantial overhang at each end of the beam. If the overhang is small, the values in the table may be slightly in error.

Before embarking on the complexities of thick-face sandwich analysis, it is prudent for the designer to check whether or not such an analysis is really necessary. Some very simple rules for doing this are given in section 8.3. It is always worthwhile to do these checks before beginning any detailed calculations for a new type of sandwich beam, column or panel.

8.3 OVERVIEW OF REGIMES OF SANDWICH BEHAVIOUR

8.3.1 The Problem

In section 8.2 sandwich panels were classified according to the three types of analysis required:

Composite beam (no shear strains)
Thin-face sandwich
Thick-face sandwich

Thick-face sandwich beams are much more complicated to analyse that thin-face sandwich beams. It is therefore natural to ask:

a) Are the differences between the results of thin-face and thick-face analyses are ever really significant?

a) What are the essential characteristics of thick-face sandwich behaviour in practice?

c) What objective criteria can be used to decide whether a sandwich is best treated as thin-face or thick-face?

Fortunately it turns out that the answer to question c) is very simple, and it can be applied without modification to beams, columns and panels.

But first, how can the answers to questions a) and b) be determined?

8.3.2 Characteristic Features of Thick-Face Sandwich Behaviour
Significant differences between thin and thick-face behaviour may be expected in cases where at least one of the following conditions exists:

(a) The value of D_f (the combined flexural rigidity of the faces) is a relatively high proportion of the total flexural rigidity D. For sandwiches with flat faces this condition is likely to be met only when the thickness of one or both of the faces is a significant fraction of the thickness of the core (20% say). But in sandwiches with corrugated faces D_f can be significantly large (10% or more of the value of D).

(b) The core shear stiffness is exceptionally low in comparison with Young's modulus for the faces. This can happen when the designer is carried away with enthusiasm for extremely low-density cores.

(c) The span of the beam (or the length of the column, or the width of the panel) is a relatively low multiple of the core thickness.

In order to understand the consequences of thick-face sandwich theory, consider the diagram in figure 8.8, which shows the primary and secondary deformations of a simply-supported beam with a central point load. The primary displacement is not quite the same as the 'bending deflection' in a thin-face sandwich, because it resists only a part of the applied load. To understand the secondary displacement, compare it with figure 8.5c, which shows the shear displacement in a thin-face beam. The most obvious difference is that the resistance of the faces to bending causes the discontinuities of gradient at the points aa, bb and cc to be smoothed out. This is what causes the secondary bending in the faces, and leads to high local

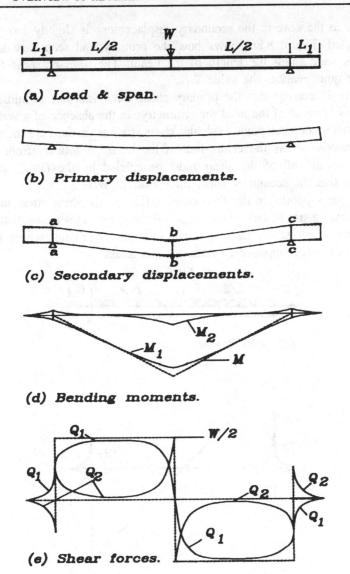

(a) Load & span.

(b) Primary displacements.

(c) Secondary displacements.

(d) Bending moments.

(e) Shear forces.

Figure 8.8. Primary and secondary displacements, bending moments and shear forces for a simply supported beam under point load.

bending stresses at the marked points. Figure 8.8d shows diagrammatically how the primary and secondary bending moments vary along the beam; the total is equal to the usual bending moment diagram.

One more point worth noting is that, since the applied load is shared between the primary and secondary displacements, the shear stress in the core in the secondary displacement is slightly less than expected. Figure 8.8e shows how the primary and secondary shear forces vary along the length of the beam. The primary shear force never quite reaches the value W/2.

It is interesting that the primary shear force vanishes at midspan. This is because of the need for symmetry; in the absence of a vertical stiffener under the point load, the shear stresses in the core must be continuous across the centre-line of the beam. It follows from this that, locally, all of the shear must be carried by the faces, which means that the secondary shear force rises to W/2.

At the supports, in the absence of stiffeners, the shear stress in the core must again be continuous across the support. This means that the primary shear stress must continue into the overhang, where it is balanced by an opposite secondary shear stress.

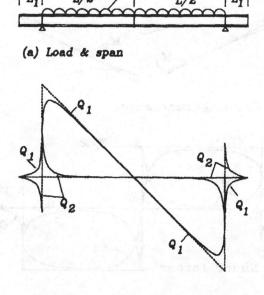

(a) Load & span

(b) Primary & secondary shear

forces diagrams.

Figure 8.9. Primary and secondary shear forces acting on a simply supported beam under uniformly distributed load.

Figure 8.9 shows the primary and secondary shear force diagrams for the case of a simply-supported beam with a uniformly-distributed load.

8.3.3 Objective Criteria for Distinguishing Between Thin- and Thick-Face Sandwich Beams, Columns and Panels

A) Sinusoidally-loaded beams

The analysis begins with the differential equations (8.17) and (8.18) for primary and secondary displacements of beam-columns. In general the displacements of a simply-supported beam-column can be regarded as the sum of a Fourier sine series, of which the first term is usually dominant. Therefore, in order to reach some general conclusions about sandwich behaviour, it will be sufficient to consider the case of a simply-supported beam column carrying sinusoidally-distributed transverse load and deforming in the shape of a sine curve.

Suppose the intensity of the transverse load is given by

$$q = q_0 \sin\left(\frac{\pi x}{L}\right) \tag{8.33}$$

where q_0 is the load amplitude and x is measured along the length of the beam. The resulting primary and secondary displacements are

$$w_1 = w_{10} \sin\left(\frac{\pi x}{L}\right) \tag{8.34}$$

$$w_2 = w_{20} \sin\left(\frac{\pi x}{L}\right). \tag{8.35}$$

If these expressions are substituted in the differential equations (8.17) and (8.18) it is fairly easy to show that all the terms involving $\sin(\pi x/L)$ cancel.

The amplitudes of the primary and secondary displacements reduce to:

$$w_{10} = -\frac{q_0 \, D_Q \, L^2}{\alpha \pi^2}; \quad w_{20} = -\frac{q_0 \, (D - D_f)}{\alpha} \tag{8.36}$$

where

$$\alpha = -D_f(D - D_f)\left(\frac{\pi}{L}\right)^4 + \left[P\,(D - D_f) - DD_Q\right]\left(\frac{\pi^2}{L}\right) + PD_Q. \tag{8.37}$$

In the case of a beam with a sinusoidal transverse load, the end-load P is zero and the total displacement can be found. It is useful to express this in terms of the ordinary bending displacement of the beam (in the absence of any shear deformation whatsoever).

Thus:

$$r = \frac{w_{10} + w_{20}}{w_0} \tag{8.38}$$

where

w_{10} is the primary displacement at midspan,
w_{20} is the secondary displacement at midspan,
w_0 is the ordinary bending displacement, ignoring shear deformation, and the ratio r is

$$r = \frac{\left[1 - \dfrac{D_f}{D} + \dfrac{D_Q}{D}\left(\dfrac{L^2}{\pi^2} \right) \right]}{\left[\dfrac{D_f}{D}\left(1 - \dfrac{D_f}{D} \right) + \dfrac{D_Q}{D}\left(\dfrac{L^2}{\pi^2} \right) \right]}. \tag{8.39}$$

In figure 8.10 the value of r is plotted for various values of (D_f/D) against $(D/D_Q L^2)$. (For sandwich beams with flat faces, each value of D_f/D corresponds to a value of c/t, which is also shown). This diagram shows how much a sandwich beam deflects in comparison with a simple composite beam with no shear deformation. It is interesting that almost identical graphs can be prepared for beams with central point load, or with uniformly distributed load.

B) Beams in general

Figure 8.10 is a master diagram with some very useful properties.
For any sandwich beam it is always easy to work out the sandwich parameters (D_f/D) and $(D/D_Q L^2)$, which fix a point on the diagram.

i)   If the point lies at the extreme left of the diagram, where r is less than (say) 1.1, the core shear deformation is small and ordinary composite beam analysis is applicable with little error.
ii) If the point lies in the region AB, or close to the line EF, the sandwich acts as a thin-face sandwich. An indication of the extent of the shear deformation can be obtained by reading the value of r from the vertical scale on the left.
iii) If the point lies to the right of the line EF (i.e. on one of the

curves on the right of the diagram), then the sandwich acts as a thick-face sandwich.

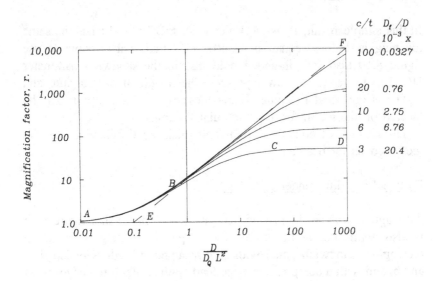

Figure 8.10. Sandwich panel master diagram.

iv) If the point lies on the nearly-horizontal part of one of the curves on the right-hand side of the diagram (such as CD) then the sandwich not only acts as a thick-face sandwich, but the core is so flexible that the faces act virtually as two independent beams. Such a structure has lost nearly all of the sandwich action, and it is likely to be very inefficient.

C) Columns

In the case of a pin-ended column, the sinusoidal primary and secondary displacements equations (8.34) and (8.35) can again be substituted in the differential equations (8.17) and (8.18). This time, however, the intensity of the transverse load (q) is zero. It is easy to show that the equations are valid only when α vanishes, in which case the right-hand side of equation (8.37) is zero. This makes it possible to obtain an expression for P, which is the critical value of the end load:

$$P_{cr} = \frac{P_E}{r}. \qquad (8.40)$$

In this simple result, P_E is the Euler load, $\pi^2 D/L^2$, and r has the same meaning as before (equation (8.39)). This makes it possible to use figure 8.10 also for pin-ended columns. In the sandwich parameter $D/D_Q L^2$, the length L now represents the length of the column, and D_f/D has the usual meaning. Using these two values, a point can be located on the graph for any particular column.

The regions of composite thin- and thick-faced behaviour can be identified as for beams.

D) Panel buckling under edge load

Although it will not be proved here, it can be shown that figure 8.10 is also applicable to the basic case of panel buckling. Consider a rectangular sandwich panel with simply-supported edges of length a and b, and with a compressive edge load applied along the edges b (in a direction parallel with the edges a).

Provided the panel is squarish, so that it buckles in the fundamental mode with one half-wave in each direction, the conventional critical edge stress must be multiplied a by a factor 1/r, where r is given by equation (8.39), except that L must now be interpreted as the width, b.

If the panel is very short in the direction of the load (a << b) then the panel buckles as a column and L should be taken as the length a and not the width b.

E) Sandwich structures of all types

Evidently figure 8.10 can be used in a wide range of circumstances to determine whether a sandwich is likely to behave as an ordinary composite beam, as a thin-face sandwich or as a thick-face sandwich. It is simply a matter of choosing the correct dimension L to use in the sandwich parameter $D/D_Q L^2$. For simply-supported beams it is the span; for columns it is the length; for squarish panels with edge loading it is normally the width perpendicular to the direction of the load; for panels with transverse pressure it is the shorter edge length.

Figure 8.10 is used to classify sandwich panels. Therefore, in cases where simple boundary conditions do not apply, some engineering judgement can be used to estimate the correct value of the characteristic length L. In the case of a fixed-ended column it would be sensible to take L as the conventional effective length, which is half the length of the column.

8.4 SANDWICH PANELS (PLATES)

8.4.1 General Equations

It is possible to obtain two sixth-order differential equations for the primary and secondary displacements of a rectangular sandwich panel, as shown in figure 8.11, for the rather general case in which:

- there is a transverse pressure of intensity q,
- there are edge forces N_x, N_y and N_{xy} (per unit length),
- the faces are thick,
- the faces are orthotropic,
- and the core is orthotropic also.

(Equations (8.17) and (8.18), for a beam-column, are special cases of these general equations).

These differential equations are given in section 8.5 for reference. For the present purpose it is more convenient to concentrate on the single differential equation for the total displacement, w:

$$\left[\nabla_{1f} \left(\nabla_1 - \nabla_{1f} \right) - \nabla_1 \nabla_3 \right] w = \left[\nabla_1 - \nabla_{1f} - \nabla_3 \right] \left(q + \nabla_2 w \right). \qquad (8.41)$$

(This equation is obtained as equation (8.77) in section 8.5). The symbols ∇ represent various differential operators, all of which are listed in section 8.5, where they are defined in terms of various flexural rigidities of the panel and its faces (D_x, D_y, D_{xf}, D_{yf}, etc.) together with panel edge forces (N_x, N_y, N_{xy}).

There is little point in attempting to solve this equation for the most general case. Instead, it will be shown how solutions can be obtained for cases of common interest. Even so, there are so many different cases that it is impossible in the space available to do more than to hint at the possibilities which exist for particular applications.

The first simplification is to assume that the faces are thin. This is probably true for the vast majority of sandwich panels. (Except for

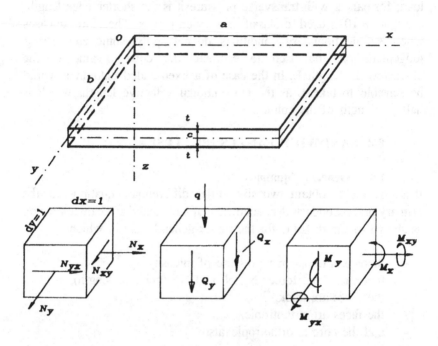

Figure 8.11. Notation for sandwich panels (plates).

panels with one face corrugated; such panels may often be more easily treated as beams spanning in the direction of the corrugations). For thin-face panels, the operator ∇_{1f} is negligible, since the flexural rigidities of the faces (D_{xf}, D_{yf}, D_{yf}) are small compared to the overall sandwich values (D_x, D_y).

This leaves:

$$- \nabla_1 \nabla_3 w = \left[\nabla_1 - \nabla_3\right]\left(q + \nabla_2 w\right). \tag{8.42}$$

In order to keep the algebra to a minimum, consider the situation in which the displacements and the transverse pressure are sinusoidal:

$$w = \sum \sum w_{ij} \sin\left(\frac{i\pi x}{a}\right) \sin\left(\frac{j\pi y}{b}\right) \tag{8.43}$$

$$q = \sum \sum q_{ij} \sin\left(\frac{i\pi x}{a}\right) \sin\left(\frac{j\pi y}{b}\right). \tag{8.44}$$

With these expressions in the differential equation, the term ij can generally be treated independently of the other terms, and the double-sine terms can be cancelled. Such a simple solution is possible only when the boundaries are simply supported and the N_{xy} term is zero (no edgewise shear force).

The equations will be more compact if we use the notation:

$$\alpha = \left(\frac{i\pi}{a}\right) \; ; \; \beta = \left(\frac{j\pi}{b}\right). \tag{8.45}$$

Note that α and β have the dimensions of [Length]$^{-1}$.

8.4.2 Case: Transverse Pressure Only; All Orthotropic; N = 0

The relationship between the amplitude of the displacement in mode ij and the amplitude of the corresponding pressure is:

$$\frac{w_{ij}}{q_{ij}} = \left(\frac{1}{\left(D_x\alpha^4 + D_1\alpha^2\beta^2 + D_y\beta^4\right)}\right)\left(1 + \frac{\left(D_x\alpha^4 + D_1\alpha^2\beta^2 + D_y\beta^4\right)}{D_{Qx}\alpha^2 + D_{Qy}\beta^2}\right).$$

$$\tag{8.46}$$

(See section 8.5 for definitions of the quantities D). For the isotropic case this reduces to:

$$\frac{w_{ij}}{q_{ij}} = \frac{1}{D\left(\alpha^2+\beta^2\right)^2}\left(1 + \frac{D\left(\alpha^2+\beta^2\right)}{D_Q}\right). \tag{8.47}$$

When the core is stiff in shear, so that D_Q approaches infinity, this reduces further to:

$$\frac{w_{ij}}{q_{ij}} = \frac{1}{D\left(\dfrac{i^2\pi^2}{a^2} + \dfrac{j^2\pi^2}{b^2}\right)^2}. \tag{8.48}$$

This is the standard result for a homogeneous isotropic plate.

It is fairly easy to recognise the significance of the terms in equations (8.46) and (8.47). The term $(D_x\alpha^4 + D_1\alpha^2\beta^2 + D_y\beta^4)$ takes the place of the usual term for the flexural rigidity of the plate, D, except that it incorporates allowances for the orthotropy of the faces. The ratio on the right hand side of equation (8.46),

$$\frac{\left(D_x\alpha^4 + D_1\alpha^2\beta^2 + D_y\beta^4\right)}{D_{Qx}\alpha^2 + D_{Qy}\beta^2},$$

is a generalisation of the sandwich parameter $D/(D_Q L^2)$, and has the same dimensions.

8.4.3 Case: Edgewise Force Nx Only; All Orthotropic; q = 0

The critical value of N_x, the edgewise force per unit length, applied in the x-direction, is obtained by writing q=0 in equation (8.42). If the equation (8.43) is substituted for w, the condition for non-zero displacement w can be shown to be:

$$N_x = -\left(\frac{\left(D_x \alpha^4 + D_1 \alpha^2 \beta^2 + D_y \beta^4\right)}{\alpha^2}\right) \bigg/ \left(1 + \frac{\left(D_x \alpha^4 + D_1 \alpha^2 \beta^2 + D_y \beta^4\right)}{D_{Qx} \alpha^2 + D_{Qy} \beta^2}\right).$$

$$(8.49)$$

For the isotropic case, this reduces to:

$$N_x = -D \frac{(\alpha^2 + \beta^2)^2}{\alpha^2} \bigg/ \left(1 + \frac{D(\alpha^2 + \beta^2)}{D_Q}\right) \tag{8.50}$$

or

$$N_x = -D \beta^2 \left(\frac{\alpha}{\beta} + \frac{\beta}{\alpha}\right)^2 \bigg/ \left(1 + \frac{D(\alpha^2 + \beta^2)}{D_Q}\right). \tag{8.51}$$

When the core is stiff in shear, so that D_Q approaches infinity, this reduces to:

$$N_x = -D \left(\frac{j\pi}{b}\right)^2 \left(\frac{i}{j}\frac{b}{a} + \frac{j}{i}\frac{a}{b}\right)^2. \tag{8.52}$$

This is the standard result for homogeneous plates. The minimum value of N_x is obtained when there is one half-wave across the width of the plate (j = 1). If the plate is square, so that there is also one half-wavelength in the x-direction, then i = 1 also:

$$N_x = -\frac{4D\pi^2}{b^2}. \tag{8.53}$$

This is another standard result.

As before, the term $(D_x \alpha^4 + D_1 \alpha^2 \beta^2 + D_y \beta^4)$ takes the place of the usual term for the flexural rigidity of the plate, D, except that it incorporates allowances for orthotropy of the faces.

The ratio on the right hand side

$$\frac{D_x \alpha^4 + D_1 \alpha^2 \beta^2 + D_y \beta^4}{D_{Qx} \alpha^2 + D_{Qy} \beta^2}$$

is again a generalisation of the sandwich parameter $D/(D_Q L^2)$ and has the same dimensions.

The negative sign for N_x arises because tension was taken as positive.

8.5 SANDWICH PANEL (PLATE) EQUATIONS

This concluding section is devoted to a brief outline of the underlying equations for sandwich panels. It is intended as starting point for anyone wishing to develop them for particular applications.

The notation and sign conventions used here are the same as those used by Timoshenko [4] (figure 8.11). The notation of Libove and Batdorf [5], used in [1], is slightly different. The properties of a sandwich panel can be defined as follows.

The bending moments M_x, M_y and the curvatures $\partial^2 w/\partial x^2$ and $\partial^2 w/\partial y^2$ are

$$M_x = - D_x \frac{\partial^2 w_1}{\partial x^2} - \upsilon_y D_x \frac{\partial^2 w_1}{\partial y^2} \qquad (8.54)$$

$$M_y = - D_y \frac{\partial^2 w_1}{\partial y^2} - \upsilon_x D_y \frac{\partial^2 w_1}{\partial x^2}. \qquad (8.55)$$

These define the flexural rigidities D_x, D_y in cylindrical bending in the zx and yz planes. The Poisson's ratios, υ_x, υ_y and the flexural rigidities D_x, D_y, etc. can be built up for given laminate types as described in the Chapter on the mechanics of orthotropic laminae. It can be shown that

$$\upsilon_y D_x = \upsilon_x D_y. \qquad (8.56)$$

The relationship between the rate of twist and the twisting moment M_{xy} is:

$$M_{xy} = 2 D_{xy} \frac{\partial^2 w_1}{\partial x \, \partial y} \qquad (8.57)$$

where D_{xy} is the torsional rigidity.

Shear stiffnesses D_{Qx}, D_{Qy} are defined in the same way as for sandwich beams. For example, D_{Qx} is the shear force needed to produce a unit gradient in the secondary deformation in the zx plane:

$$Q_x = D_{Qx} \frac{\partial w_2}{\partial x}, \qquad Q_y = D_{Qy} \frac{\partial w_2}{\partial y}. \tag{8.58}$$

Thus there are seven basic quantities (D_x, D_y, v_x, v_y, D_{xy}, D_{Qx}, D_{Qy}) of which only six are independent in general. However, it is necessary to distinguish between D_x, D_y and D_{xy} (which refer to the complete sandwich) and D_{xf}, D_{yf}, D_{xyf} (which represent the combined stiffnesses of the faces when they act as two independent thin plates). It is also convenient to write:

$$D_{x0} = D_x - D_{xf}; \qquad D_{y0} = D_y - D_{yf}; \qquad D_{xy0} = D_{xy} - D_{xyf} \tag{8.59}$$

The quantities D_{x0}, D_{y0}, D_{xy0} are the stiffnesses of the panel when only uniform membrane stresses in the faces are considered.

The various components of the potential energy of a panel with transverse and edge loads are as follows.

The strain energy of the faces, due to primary deformation, considering only axial (membrane) stresses in the faces, is:

$$U = \int_0^b \int_0^a \left[\frac{D_{x0}}{2} \left(\frac{\partial^2 w_1}{\partial x^2} \right)^2 + \frac{D_{y0}}{2} \left(\frac{\partial^2 w_1}{\partial y^2} \right)^2 + \right.$$

$$\left. \left(\frac{v_y D_{x0} + v_x D_{y0}}{2} \right) \frac{\partial^2 w_1}{\partial x^2} \frac{\partial^2 w_1}{\partial y^2} + 2 D_{xy0} \left(\frac{\partial^2 w_1}{\partial x \, \partial y} \right)^2 \right] dx \, dy. \tag{8.60}$$

The strain energy of the core, due to secondary deformation, is:

$$U = \int_0^b \int_0^a \left[\frac{D_{Qx}}{2} \left(\frac{\partial w_2}{\partial x} \right)^2 + \frac{D_{Qy}}{2} \left(\frac{\partial w_2}{\partial y} \right)^2 \right] dx \, dy. \tag{8.61}$$

The strain energy of the faces due to local bending and twisting in the primary and secondary deformations ($w = w_1 + w_2$) is:

$$U = \int_0^b \int_0^a \left[\frac{D_{xf}}{2} \left(\frac{\partial^2 w}{\partial x^2} \right)^2 + \frac{D_{yf}}{2} \left(\frac{\partial^2 w}{\partial y^2} \right)^2 + \right.$$

$$\left. \left(\frac{\nu_y D_{xf} + \nu_x D_{yf}}{2} \right) \frac{\partial^2 w}{\partial x^2} \frac{\partial^2 w}{\partial y^2} + 2 D_{xyf} \left(\frac{\partial^2 w}{\partial x \partial y} \right)^2 \right] dx\, dy. \qquad (8.62)$$

The potential energy of position of the transverse load of intensity q is:

$$V_p = - \int_0^b \int_0^a q\, w\, dx\, dy. \qquad (8.63)$$

The potential energy of position of the edge forces of intensity Nx, Ny, Nxy is:

$$V_p = \int_0^b \int_0^a \left[\frac{N_x}{2} \left(\frac{\partial w}{\partial x} \right)^2 + \frac{N_y}{2} \left(\frac{\partial w}{\partial y} \right)^2 + N_{xy} \left(\frac{\partial w}{\partial x} \frac{\partial w}{\partial y} \right) \right] dx\, dy. \qquad (8.64)$$

The total potential energy, V, is the sum of the expressions in equations (8.60) to (8.64); it can be written in the form:

$$V_p = \int_0^a \int_0^b F\left(w_1,\ w_2,\ w_{1x},\ w_{1y},\ w_{2x},\ w_{2y},\ w_{1xx}\ ... \right) dx\, dy; \qquad (8.65)$$

$$\left(\text{where}\quad w_{1x} = \frac{\partial w_1}{\partial x}\quad \text{etc.} \right).$$

The condition for V to be stationary with respect to small variations in the displacement function $w_1(x,y)$ is that the latter must satisfy the differential equation:

$$\frac{\partial F}{\partial w_1} - \frac{\partial}{\partial x} \frac{\partial F}{\partial w_{1x}} - \frac{\partial}{\partial y} \frac{\partial F}{\partial w_{1y}} + \frac{\partial^2}{\partial x^2} \frac{\partial F}{\partial w_{1xx}}$$

$$+ \frac{\partial^2}{\partial x \partial y} \frac{\partial F}{\partial w_{1xy}} + \frac{\partial^2}{\partial y^2} \frac{\partial F}{\partial w_{1yy}} = 0. \qquad (8.66)$$

This reduces to the following expression:

$$\nabla_1 w_1 + \nabla_{1f} w_2 = q + \nabla_2 w_1 + \nabla_2 w_2 \qquad (8.67)$$

where

$$\nabla_1 = D_x \frac{\partial^4}{\partial x^4} + D_1 \frac{\partial^4}{\partial x^2 \partial y^2} + D_y \frac{\partial^4}{\partial y^4} \qquad (8.68)$$

$$\nabla_{1f} = D_{xf} \frac{\partial^4}{\partial x^4} + D_{1f} \frac{\partial^4}{\partial x^2 \partial y^2} + D_{yf} \frac{\partial^4}{\partial y^4} \qquad (8.69)$$

$$\nabla_2 = N_x \frac{\partial^2}{\partial x^2} + 2 N_{xy} \frac{\partial^2}{\partial x \partial y} + N_y \frac{\partial^2}{\partial y^2} \qquad (8.70)$$

$$D_1 = 4 D_{xy} + \upsilon_y D_x + \upsilon_x D_y \qquad (8.71)$$

$$D_{1f} = 4 D_{xyf} + \upsilon_y D_{xf} + \upsilon_x D_{yf} \qquad (8.72)$$

An equation analogous with equation (8.66) exists for the displacement function $w_2(x,y)$, which results in the following differential equation:

$$\nabla_{1f} w_1 + \nabla_{1f} w_2 - \nabla_3 w_2 = q + \nabla_2 w_1 + \nabla_2 w_2 \qquad (8.73)$$

where

$$\nabla_3 = D_{Qx} \frac{\partial^2}{\partial x^2} + D_{Qy} \frac{\partial^2}{\partial y^2}. \qquad (8.74)$$

By suitable manipulations of equations (8.67) and (8.73) it is possible to obtain differential equations for w_1 and w_2 separately:

$$\begin{aligned} &\left[\nabla_{1f} (\nabla_1 - \nabla_{1f}) - \nabla_1 \nabla_3 \right] w_1 + \\ &\left[\nabla_3 - (\nabla_1 - \nabla_{1f}) \right] \nabla_2 w_1 = - \nabla_3 q \end{aligned} \qquad (8.75)$$

$$\begin{aligned} &\left[\nabla_{1f} (\nabla_1 - \nabla_{1f}) - \nabla_1 \nabla_3 \right] w_2 + \\ &\left[\nabla_3 - (\nabla_1 - \nabla_{1f}) \right] \nabla_2 w_2 = + (\nabla_1 - \nabla_{1f}) q. \end{aligned} \qquad (8.76)$$

These equations may be added together to provide a single differential equation for the total displacement, $w = w_1 + w_2$:

$$\left[\nabla_{1}(\nabla_{1} - \nabla_{1'}) - \nabla_{1} \nabla_{3} \right] w =$$

$$\left[(\nabla_{1} - \nabla_{1'}) - \nabla_{3} \right] (q + \nabla_{2} w). \tag{8.77}$$

A direct relationship between w_1 and w_2 is obtained by subtracting equation (8.73) from (8.67):

$$\nabla_3 w_2 = -(\nabla_1 - \nabla_{1'}) w_1. \tag{8.78}$$

For the case of an isotropic panel, the differential equation (8.77) reduces to

$$\nabla^2 (D D_Q \nabla^4 - D_I D_0 \nabla^6) w =$$

$$\nabla^2 (D_Q - D_0 \nabla^2) \left[q + N_x \frac{\partial^2 w}{\partial x^2} + 2 N_{xy} \frac{\partial^2 w}{\partial x \partial y} + N_y \frac{\partial^2 w}{\partial y^2} \right] \tag{8.79}$$

where

$$\nabla^2 = \frac{\partial^2}{\partial x^2} + \frac{\partial^2}{\partial y^2}, \qquad \nabla^4 = \nabla^2 \nabla^2 \text{ etc.}$$

Apart from the extra operator ∇^2 on each side, this is the result obtained by Reissner in 1948.

8.6 REFERENCES

1] Allen, H.G., "Analysis & Design of Structural Sandwich Panels", Pergamon, Oxford, 1969.

2] Allen, H.G., "Sandwich Construction", Civ. Engg. Dept. Report CE/1/72, Southampton University, 1972.

3] Allen, H.G., "Sandwich Panels with Thick or Flexurally Stiff Faces", Proc. Conf. *Sheet Steel in Building*, Iron & Steel Inst. and RIBA, London, March, 1972. pp 10-18.

4] Timoshenko, S., "Theory of Elastic Stability", various editions, McGraw Hill, New York, 1976.

5] Libove, C., Batdorf, S.B., "A General Small-Deflection Theory for Flat Sandwich Plates", NACA TN 1526, 1948. (Also NACA Report 899)

9 DESIGN OF ANISOTROPIC PANELS

9.1 INTRODUCTION

Design of any artefact is not an activity undertaken in isolation. Although the designer may start, metaphorically, with a blank sheet of paper, he will have a solution in mind based usually on historical or traditional perceptions. There will also be in mind the function or purpose of the artefact which is the reason for the design. Ideally it should be possible to start with the function and to derive the required characteristics of the artefact directly. However, this direction of attack will only be possible for the simplest of objects, and for all practical cases it will be necessary to conceive characteristics and to predict the function from them. If the function does not match that which is required then an iterative procedure is followed.

It is also necessary to be aware of the environment in which the artefact is to function. A description of the environment in terms of loads, temperature, humidity, etc. and the time variation of these will usually be part of the functional specification. If not, it must be derived by the designer before he can start on the design.

Moving on from the philosophy, what is required for the design of plate panels? In the present context a plate panel will be part of a marine structure, and will be loaded by some or all of in-plane biaxial loads, lateral loads and shear, and these loads will frequently be cyclic. The environment is likely to be wet, and in some parts of the structure may well be hot.

In meeting the requirement the designer will usually start with the concept of an isotropic material, although the traditional marine structural solution will use that material in an orthotropically stiffened manner. Indeed the stiffened flat steel plate solution is difficult to beat. It can be made efficient for both in-plane and lateral loads, as well as being resistant to shear. Its principal disadvantages in the marine environment are its density, its susceptibility to corrosion and the residual stresses and distortion that are a result of welded fabrication.

Why then should the designer be considering anisotropic materials?

236

The answer is that generally he will not, but he will be trying to overcome the disadvantages of steel construction and will be driven to composite materials, usually some form of fibre reinforced plastic (FRP). While strictly anisotropic the response of FRP is primarily orthotropic and so in the first instance it can be treated in a similar way to a stiffened isotropic panel.

In practice, therefore, the design procedure commences with consideration of an orthotropic panel without an immediate need to decide how that orthotropy is to be provided. In the case of FRP it may be by using an unbalanced laminate or by using some form of unidirectional or bidirectional stiffening or by a combination of both. The design procedure as presented here will be subdivided by load direction (lateral, in-plane, etc.) and then developed to show how best to use anisotropic materials.

9.2 LOADING AND LOAD FACTORS

As indicated above, one of the first things to be considered is the loading to be expected on the panel, and the variability of the load. In most marine structures it will be possible to predict a maximum load, although this will be subject to uncertainties in the form of partial safety factors, by which the maximum load is multiplied. For the purposes of this Chapter, it will therefore be assumed that a maximum design load in each of the modes in question has been identified, and that no further load factors are needed.

Cyclic load will lead to considerations of fatigue failure. It is not the purpose of this Chapter to consider the fatigue analysis of composite materials which is to be covered later, but suffice it to say that it is conventional practice to undertake design of panels neglecting fatigue considerations except to keep stresses reasonably low if there is a significant cyclic content, and to analyse for fatigue after the design is complete, iterating further if fatigue criteria are not met.

A load or strength factor will be needed to allow for the effect of water absorption in FRP over the life of the structure. Generally it will be necessary to obtain advice from the material manufacturers, but as a guide it has been shown [1] that for glass reinforced polyester resin both the mean strength and stiffness reduce by between 10% and 20% over a number of years in service.

Creep, while not a load but a response to load applied continuously for a long period, is a phenomena that rarely needs to be considered in design of panels for marine structures. This is because the large majority of loads are cyclic and so a creep response can never occur. However, it may be necessary to provide margins against creep when loading is due to the mass of heavy items of equipment, especially if the structure is at a higher than ambient temperature, such as in a machinery space. A discussion of creep is covered in standard text books, for example reference [1].

9.3 DESIGN OF PANELS UNDER LATERAL LOAD

9.3.1 Design Principles

Although this book is concerned with composite materials, an example of the principles of design can be gleaned from metal structures. Metal itself of course can be treated as isotropic, but orthogonally stiffened panels are clearly orthotropic and so a discussion of the design processes for stiffened metal panels provides a good guide into the design of anisotropic panels.

Plating thickness and stiffener spacing will usually be decided with reference to adjacent structure and using simple isotropic plate theory. The grillage is also likely to have one set of stiffeners significantly stiffer than the other set, as illustrated in figure 9.1. In such a case a

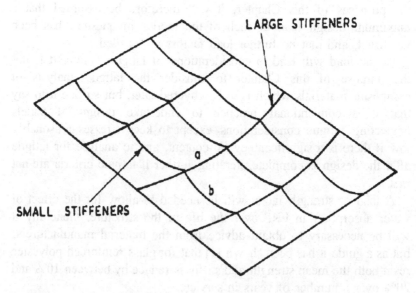

Figure 9.1. Typical arrangement of large and small stiffeners on a grillage.

sufficiently reliable start to the design can be made by sizing the stiff beams using simple beam bending theory and neglecting the less stiff beams. This will be done taking account of the actual load and boundary conditions and an appropriate approximation to plate effective breadth. A suitable stress factor (say 1.2 on yield) will then lead to the scantlings of the stiffener.

The scantlings of the less stiff beams are not quite so easily derived because the deformation of the stiffer beams leads to differential deflections between the ends of the element of less stiff beams between stiffer beams. It is therefore necessary to make an estimate of the worst relative deflection along an element of the less stiff beam, usually at the centre of an edge. With this relative deflection, and again with an appropriate estimate for effective plate breadth, but this time using clamped ends (as a conservative assumption) it is a simple matter to size the less stiff beams to withstand the lateral pressure. Scantlings are then refined using orthotropic theory as given in standard text books.

Leading on to composite panels it is clear that a similar approach could be used. That is, given a particular combination of reinforcement and matrix whose flexural properties are known, only the thickness then needs to be derived. The higher stress will occur in the short direction across the panel and so a strip of unit width with the appropriate boundary conditions can be treated as a simple beam and the thickness calculated. In this case there is no other parameter to be found and stresses in the long direction will automatically be lower.

However, while elastic design procedures are used both for metals and composites, it must not be forgotten that the composite material is essentially anisotropic, although in marine applications the fibres will almost invariably be either random or aligned parallel to the panel edges so orthotropic theory is adequate. A more significant problem is the low stiffness of marine composites relative to metals, and so panels are likely to be deflection limited rather than stress limited and even then may need consideration of large deflection theory.

The same general principles will apply to sandwich panels and to rib stiffened composite panels, where in the latter case a combination of composite plate design and orthogonally stiffened metal structure design procedures may be used. Procedures are discussed in more detail in the following sections.

9.3.2 Design of Single Skin Panels

As a first approximation to the design the anisotropy of the material may be ignored and an average value of Young's modulus (E) and Poisson's ratio (v) taken from material property tables. In this case, for a rectangular flat plate of uniform thickness under uniform pressure (P) the maximum deflection (w) is given by:

$$w = \alpha P a^3 b \frac{(1 - v^2)}{E h^3} \tag{9.1}$$

and the maximum flexural stress (σ) is:

$$\sigma = cP \frac{ab}{h^2} \tag{9.2}$$

α and c are non-dimensional parameters depending on aspect ratio and boundary conditions. Tables for α and c are given in reference [2] and in a similar form in reference [3] except that the latter has to be modified for Poisson's ratio which is there assumed to be 0.3. The references also discuss point and patch loading and provide the appropriate formulae and constants.

It is clearly a straightforward matter, provided an allowable value of stress, see below, or allowable deflection is specified, and knowing the material constants, to calculate an appropriate value of h. This is unlikely however to match an integer number of laminates and will need to be rounded up accordingly.

It is possible that the allowable deflections will be greater than that appropriate to small deflection theory, that is if $w/h > 0.5$. In such a case membrane action becomes significant and the plate will be stiffer than predicted by the formulae above. Modified formulae are as follows, and clearly must be solved in an iterative numerical manner

$$P = \frac{E h^4}{a^3 b(1 - v^2)} \left[\frac{1}{\alpha}\left(\frac{w}{h}\right) + \gamma\left(\frac{w}{h}\right)^3 \right] \tag{9.3}$$

$$\sigma = \frac{E h^2}{a^2} \left[\frac{c}{\alpha(1 - v^2)} \left(\frac{w}{h}\right) + \alpha_1\left(\frac{w}{h}\right)^2 \right] \tag{9.4}$$

There is very limited data available for the values of the non-dimensional parameters γ and α_1 but for practical purposes a value of $\delta = 2$ and $\alpha_1 = 1.6$ (simply supported) or 2.9 (clamped) are recommended.

The formulae given above are derived from isotropic considerations and are suitable for first estimates or if the laminate is a balanced one (that is the flexural modulus is the same in both principal directions). If the laminate contains unidirectional fibres or unbalanced woven rovings then there will be two values of Young's Modulus E_1 and E_2 (in x and y directions, assumed $E_1 > E_2$) and a Poisson's Ratio v_{12}. In this case stiffness parameters are defined as

$$D_1 = \frac{E_1 h^3}{12\mu} \text{ and } D_2 = \frac{E_2 h^3}{12\mu} \qquad (9.5)$$

where $\mu = 1 - v_{12}^2 \dfrac{E_2}{E_1}$.

The maximum deflection is then given by

$$w = \alpha P \frac{a^3 b \mu}{E_2 h^3} \qquad (9.6)$$

and maximum stresses in the two directions are given by

$$\sigma_1 = c_1 \frac{Pab}{h^2} \text{ and } \sigma_2 = c_2 \frac{Pab}{h^2}. \qquad (9.7)$$

Values of α, c_1 and c_2 depend on D_1/D_2 as well as aspect ratio and can be found in reference [4].

Material data for input to these formulae is clearly dependant on lay-up and fibre-volume fraction and in most cases will need to be calculated using a method such as discussed in Chapter 5 or by means of mechanical testing. Some guidance on properties of commonly used shipbuilding materials is given in references [1] and [5].

Where the limiting parameter is stress and not deflection, consideration needs to be given to the allowable maximum stress value. For practical purposes FRP acts elastically up to failure so no allowance needs to be made for plasticity. Failure is usually defined as First-Ply-Failure (FPF) and in the case of flexure usually manifests itself as cracking or crazing of the surface resin. This should be avoided in marine structures because such cracks allow ingress of water to the laminate with consequent degradation of strength and stiffness. Failure mechanisms are discussed in Chapter 11 but as a first approximation Smith [1], recommends that the design allowable strain (or stress) is not more than 30 per cent of the ultimate failure strain (or stress) and preferably nearer 25 per cent.

9.3.3 Design of Sandwich Panels

The sandwich beam and panel theory has been covered in depth elsewhere in Chapter 8. Although sandwich structures may be defined as having low shear stiffness or high shear stiffness cores and each will react differently, for practical purposes only high shear stiffness cores are suitable for marine construction, that is those with end-grain balsa or rigid PVC foam, or similar. If the cores are sufficiently thin for shear to be ignored that is if $h_c/h \not> 0.5$ then the following procedure is valid; otherwise the more complex procedure outlined below is relevant.

If skins have suffix 'S' and cores suffix 'C' then using the rule-of-mixtures the tensile modulus of the sandwich is given by

$$E_T = E_C \left(\frac{h_c}{h}\right) + E_S \left(1 - \frac{h_c}{h}\right) \tag{9.8}$$

while the flexure modulus is given by

$$E_F = E_C \left(\frac{h_c}{h}\right)^3 + E_S \left(1 - \left\{\frac{h_c}{h}\right\}^3\right) \tag{9.9}$$

Note that the tensile and flexural moduli are different and $E_F > E_T$ if $E_S > E_C$.

Having determined the value of E_F the flexural rigidity can be found in the usual way

$$D = \frac{h^3 E_F}{12}. \tag{9.10}$$

Considering the strength of the sandwich there are two elastic failure modes, tension or compression failure of the skin or tension failure of the core at the core skin interface. If σ_{sf} is the failure stress of the skin material and σ_{cf} is the failure stress of the core, then the limiting bending moments are given by

$$M_f = \frac{h^2 E_F \sigma_{sf}}{6 E_s}. \tag{9.11a}$$

and

$$M_f = \frac{h^3 E_F \sigma_{cf}}{6 h_c E_C} \tag{14.11b}$$

For an applied pressure (P) the maximum bending moment (M) and deflection (w) are given by

$$M = \frac{cPab}{6} \qquad (9.12)$$

$$w = \frac{\alpha Pa^3 b}{12D} \qquad (9.13)$$

where c and α are the same as those discussed in the previous section.

The main advantage of sandwich construction arises because the low density core material enables panel thickness (h) to be increased for the same weight. It is therefore the h^2 and h^3 terms in equation (9.11) which offset the lower flexural modulus and lead to higher stiffness and strength. However, increasing core thickness also increases weight and attention may need to be given to minimum weight design. In the stiffness limited case it is shown [2] that the optimum core to total thickness ratio is given by

$$\frac{h_c}{h} = \left[\frac{1 - \rho_c/\rho_s}{1 - E_c/E_s}\right]^{\frac{1}{2}} \qquad (9.14)$$

where P_c and P_s are the densities of the core and skin materials respectively. Using this equation E_F may be found from equation (9.9), and equation (9.10) then gives the required relationship between flexural rigidity and total thickness.

The above procedure is relevant where the core is thin and stiff, and shear in the core can be neglected. Frequently in marine sandwich panels the core is relatively thick, although of stiff material and in this case core shear must be allowed for. The detailed theory is presented by Smith [1] but for design purposes the following procedure may be used.

The flexural and shear rigidities of the panel should be calculated using the expressions below, which assume that the upper and lower skins are identical and the core carries shear load only. If the skins differ then the more complex procedure of reference [1] should be used.

$$D_1 = \tfrac{1}{2} E_{S1} h_S (h - h_S)^2 / (1 - v_1 v_2) \qquad (9.15a)$$

$$D_2 = \tfrac{1}{2} E_{S2} h_S (h - h_S)^2 / (1 - v_1 v_2) \qquad (9.15b)$$

where E_{S1} and E_{S2} are the orthotropic Young's moduli of the skins (D_2

is stiffness in short direction),

$$D_C = G_C \frac{(h - h_s)^2}{h_C}$$

(9.16)

where G_C is the shear modulus of the core.

For design purposes it is usually then sufficient to assume cylindrical bending so that the maximum deflection is given by

$$w = \frac{5Pb^4}{384D_2} + \frac{Pb^2}{8D_C} \text{ (simply supported)}$$

(9.17)

$$\text{or} \quad w = \frac{Pb^4}{384D_2} + \frac{Pb^2}{8D_C} \text{ (clamped)}$$

and maximum stress

$$\sigma = \frac{P}{8} \cdot \frac{b^2 h}{h_s (h - h_s)^2} \cdot \text{(simply supported)}$$

(9.18)

$$\text{or} \quad \sigma = \frac{P}{12} \cdot \frac{b^2 h}{h_s (h - h_s)^2} \cdot \text{(clamped)}$$

core shear stress (τ_C) is

$$\tau_c = \frac{Pb}{2(h - h_s)} \cdot$$

(9.19)

Materials and failure data for the skins will be the same as for solid FRP. Data for core materials will need to be gleaned from manufacturers publications, but as a guide reference [1] gives data for a number of candidates.

Once the initial estimation of scantlings is complete the sandwich structure must be analysed to check it against all modes of failure including core shear and skin buckling. Detailed procedures of the properties and performance of sandwich panels may be found in reference [6].

9.3.4 Design of Rib Stiffened Panels
It is frequently the case in marine structures that to provide an acceptable stiffness to weight ratio it is necessary to stiffen single skin FRP panels with ribs running unidirectionally or orthogonally.

Detailed means of analysing such stiffened panels are given in reference [1], but for design and as a first approximation two orthogonal sets of stiffeners may be assumed to act independently, that is the torsional effects at their junction are ignored. It is then possible to treat the panel in a similar way to an anisotropic laminate as described in section 9.3.2.

The stiffness parameters are now D_{11} and D_{22} given by

$$D_{11} = D_1 + \left(D_{SX} + A_{SX}\, e_x^2\right)/b'$$ (9.20a)

$$D_{22} = D_2 + \left(D_{SY} + A_{SY}\, e_Y^2\right)/a'$$ (9.20b)

where D_1 and D_2 are as defined in equation (9.5), D_{SX} and D_{SY} are the flexural rigidities of the rib stiffeners about their own centres of area, A_{SX} and A_{SY} are the areas of the stiffeners, and e_X and e_Y are the distances between their respective centres of area and the mid plane of the plate. a' is the spacing of the Y direction stiffeners and b' the spacing of the X direction stiffeners.

Table 9.1. Values of α.

D^* \ a/b	0.25	0.5	1	2	4
1	0.0385	0.0605	0.0487	0.0152	0.0024
2	0.0195	0.0363	0.0384	0.0146	0.0024
4	0.0099	0.0196	0.0278	0.0140	0.0024
7	0.0055	0.0112	0.0190	0.0124	0.0024
10	0.0039	0.0080	0.0148	0.0116	0.0024
15	0.0026	0.0054	0.0106	0.0100	0.0024

The maximum deflection (w) is then given by

$$w = \frac{\alpha P a^3 b}{12 D_{22}} \text{ if } D_{11} > D_{22}, \ D^* = D_{11}/D_{22}$$

(9.21)

or $$w = \frac{\alpha P a b^3}{12 D_{11}} \text{ if } D_{22} > D_{11}, \ D^* = D_{22}/D_{11}$$

with α taken from table 9.1.

Rib stiffened structures under lateral load are usually deflection limited rather than stress limited and so estimation of stresses at this stage is not necessary. However, once the panel has been designed the stresses must be calculated and scantlings changed if stresses are too high. Elastic stresses may be found by finite element analysis, and folded plate analysis may also be used for unidirectional stiffening, see section 9.4.2 below.

9.4 DESIGN OF PANELS UNDER IN-PLANE LOADING

9.4.1 Design Principles

In well-designed metal structures the primary failure modes for in-plane loading are associated with elasto-plastic buckling, and procedures such as load shortening curves have been developed which allow for the interaction of elastic buckling with yielding, accounting also for initial deflections and residual stresses. FRP can have much the same strength as a steel structure but its stiffness is between 10% and 15% of steel. Moreover it has no useful post buckling strength, and although residual stresses and initial deflections may exist they are small and can be neglected during the design process.

The consequence is that the failure of FRP structures under in-plane load will be by pure elastic buckling which in principle makes the design problem an easy one. For flat plates the solution is trivial and can be found in publications such as Roark [3] although adequate safety margins should be used. Smith [1] recommends a factor of 1.5-2.0 against simple Euler buckling providing that material uncertainties are accounted for by using a reduced Young's modulus. The commonly used value for material properties (see reference [5]) is the mean less two standard deviations.

However, when sandwich panels or rib stiffened panels are used, complex failure modes may be apparent [1] and any modelling must take account of those modes and be sufficiently detailed to include a full geometric description of the stiffeners. Finite element (FE) analysis may be used if a suitable programme is available that allows for non-linearities inherent in FRP elastic buckling. An alternative developed by Smith [1] and summarised below is that of folded plate analysis which is particularly valuable for unidirectionally stiffened FRP panels.

9.4.2 Folded Plate Analysis

This technique has proved particularly useful for analysis of stiffened panels under any combination of lateral and in-plane load. The panel is idealised as an array of parallel beams and interconnected flat rectangular orthotropic plates contained between and lying normal to parallel planes $X = 0$ and $X = L$. Beams and plates are of length L and are assumed to be simply supported at $X = 0,L$. Structures which may be treated in this way include panels of orthotropic plate stiffened in a single direction with a cross-section that is uniform over the span. Orthogonally stiffened structures may also be treated provided that transverse stiffeners together with the plating can be presented as equivalent orthotropic plates.

Beams in the structure are assumed to behave according to simple beam theory, plane sections remaining plane under load. Beams may be genuine structural components corresponding to the stiffeners in a plate-stiffener assembly and possessing flexural, torsional and axial stiffness. Alternatively they may be fictitious elements with zero rigidity simply acting as lines of reference for plate intersections or line loads.

The deflections, forces and moments on the beams are represented as appropriate Fourier series and are related to each other by simple beam theory. Loads applied to the beam are similarly represented by Fourier Series.

The plate elements are assumed to be governed by orthotropic plane stress equations with mid-plane symmetry. Solutions of these equations are assumed in the form of an infinite sine or cosine series where the coefficients are obtained relating to the shape of the lengthwise edge deflection. Stresses can also be defined in the same form and related to the deflections through the orthotropic plate bending equations.

Continuity conditions are then defined along the interconnecting boundaries between plates and beams, followed by force equilibrium conditions. The problem is then solved in a similar manner to that of finite element analysis, that is the matrix equation

$$K_n \, \Delta_{Bn} = - \, R_n \tag{9.22}$$

is inverted. K_n is a square stiffness matrix of order 4 x number of beams, Δ_{Bn} is a column matrix of unknown beam displacements and R_n is a column matrix of forces and moments. Separate matrices exist for each harmonic of the Fourier Series and have to be solved for each Fourier component of load and initial deflection. From the beam displacements the plate edge displacements can be found from the continuity conditions, and displacements and stresses are then readily computed at any point in each plate. Displacements and stresses are then summed for each harmonic until satisfactory convergence is found. At the same time by finding the values of in-plane forces for which the determinant of K_n is zero, the critical buckling conditions can be assessed.

9.4.3 Design of Sandwich Panels

Buckling loads and modes for a simply supported rectangular sandwich panel can be derived from the theory described by Smith [1], which is the same as mentioned in section 9.3.3. For short panels (aspect ratio less than about 3) the full theory should ideally be used, but for larger panels the Euler formula corrected for shear is acceptable, that is the critical in-plane buckling load per unit width (F_{cr}) is given by

$$F_{cr} = \frac{n^2 \pi^2 D_2}{b^2} \Bigg/ \left(1 + \frac{n^2 \pi^2 D_2}{b^2 D_c} \right) \tag{9.23}$$

where the rigidity parameters are as described in section 9.3.3. Clearly the lowest value of F_{cr} occurs where n = 1.

Solutions for clamped panels and for those involving in-plane shear are much more complicated. In such cases it is best to design using the above simply supported formula, and then to analyse and redesign as necessary with the true load and boundary conditions using FE or folded plate analysis.

In addition to the overall buckling solution described above, local instabilities of one or both faces may occur, see figure 9.2. Single sided face wrinkling is only likely to happen if one face is thinner than the other, but none of the modes shown in figure 9.2a are readily deducible from simple formulae. Smith [1] describes an approach taken from Allen [7], but in practical marine structures the skins are likely to be too thick for this type of failure. More likely may be the peeling failure shown in figure 9.2b which can occur due to local damage or imperfect fabrication. Such failure will severely degrade the strength of the panel and should be avoided by careful quality control and in-service inspection.

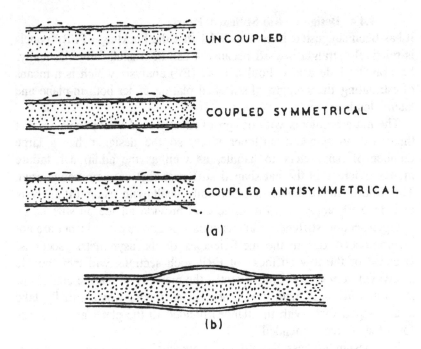

Figure 9.2. Buckling of sandwich panels: a) Face wrinkling/crimping modes, b) Skin separation.

Approximate formulae are given in reference [6] for the failure modes in figure 9.2a, which are, for one face (described as "wrinkling")

$$\sigma_{Scr} \approx 0.8 \, (e_s E_C G_C)^{\frac{1}{2}} \tag{9.24}$$

and for both faces (described as "crimping")

$$\sigma_{Scr} \approx \frac{G_C h_C}{2 h_S} \tag{9.25}$$

where σ_{Scr} is the critical skin stress.

The reliability of these formulae is not established, but they are presented as a guide.

Material properties for buckling calculations will be the same as those referenced in section 9.3.3. In view of the uncertainties involved in the calculations it is suggested that a safety factor of at least 2 is used against the various critical loads and stresses, in conjunction with worst likely material properties.

9.4.4 Design of Rib Stiffened Panels

It has been suggested in section 9.4.1 that the buckling of FRP panels is relatively straightforward because it can be treated fully elastically. Section 9.4.2 described Folded Plate (FP) analysis which is a means of estimating the strength of stiffened plating under both in-plane and lateral load.

The main problems with design of rib stiffened FRP panels are that there are no standard stiffener sizes, so the designer has a large number of parameters to handle, as well as the additional failure modes inherent in the hat shaped stiffeners that are commonly used. The failure modes however are readily dealt with using FP analysis or indeed FE analysis, so it remains to provide advice on stiffeners.

Open section stiffeners such as angles or Zs are possible but are not recommended due to the inefficiencies of the asymmetric sections. Because of the low stiffness of FRP such sections will trip (buckle sideways) very easily. Consequently the majority of marine composite structures are built with closed section stiffeners which usually take a hat shaped form with the flanges bonded to the plate, and it is this form that is recommended.

The design process therefore commences as suggested in section 9.4.1 by deducing a suitable value (or values) of b/t to avoid plate buckling between stiffeners. b and t separately may then be deduced from practical considerations, b usually being between 0.75 and two metres. If then the whole area of the stiffened panel is calculated to have a factor of safety of about five against compressive in-plane failure, and the area of plating is subtracted, the remainder must be accounted for by the stiffeners. The number of stiffeners will be clear from the spacing and so the area of each stiffener will be known.

Generally the wall and flange thickness of the stiffener is 50% to 80% of the plate thickness and the remaining material is in the table which should be about twice the thickness of the walls. The cross-section area of a stiffener will also be between 20% and 40% of the total cross-section. Using those proportions there is sufficient data to design a practical stiffener, noting that the flanges will be about 50 mm wide and the hat height between 1.5 and 2 times the width.

The first stage of the design will now be complete and it remains to analyse the arrangement to ensure compliance with design criteria. The next stage is therefore to isolate a single stiffener and effective width of plating. The effective width of steel and aluminium plating has been extensively researched and can be evaluated with reasonable accuracy, but the effective width of composite plating has been less thoroughly studied. As imperfections (initial distortions and residual stresses) are normally small the initial effective width of FRP is high, but when buckling occurs the effective width drops rapidly to between 40 per cent and 70 per cent of the full plate width [1].

Smith [1] also recommends a safety factor of 2.0 between critical buckling stress and the stress under the maximum design load because the critical stress can be predicted with reasonable accuracy. Consequently as a first estimate an effective width of 50% of stiffener spacing is recommended.

With this effective width it is then a simple matter to calculate the Euler buckling stress for the stiffener/plate combination and to compare with the actual stress in the light of the recommended safety margin above. Further analysis, to take account of other loads and any complexities in the arrangement may be carried out using FE or FP analysis.

The above discussion assumes that the panel is unidirectionally stiffened in the direction of the principal load. In practice this will be the preferred configuration unless lateral loads are dominant. It should be noted that orthogonal stiffening in FRP is difficult and expensive to achieve, especially with hat stiffness, because of the complexity of designing and fabricating the stiffener crossings.

Longitudinal stiffening is without doubt the most efficient arrangement. However, in the past for fabrication reasons it has been found necessary to stiffen panels transversely. Although such an arrangement is now likely to be rare, should it be desirable to stiffen the structure unidirectionally at right angles to the in-plane load, it will be necessary to make reference to Smith [1]. There are a number of possible failure modes involving interactions between the plate and the walls of the hat stiffeners and solutions are complex. The reference does however give guidance and some data curves for predicting the three principal types of interframe buckling with further discussion of overall buckling.

Some guidance on other composite stiffener configurations can be found in reference [8], Chapters 11 and 12.

9.5 DESIGN OF PANELS UNDER SHEAR LOADING

In a ship's side shell structure, in the webs of frames and girders and in some other parts of the structure significant shear stresses may occur. An adequate estimate of initial local buckling stress (Q_{cr}) may be made considering an infinite strip of laminate of width b with simply supported edge, using the following equation:

$$Q_{cr} = \frac{4k}{b^2 h} (D_1 D_2^3)^{¼} \qquad (9.26)$$

where k depends on the factor $\alpha = \dfrac{H}{\sqrt{D_1 D_2}}$ and when

α	=	0	0.2	0.4	0.6	0.8	1.0
k	=	8.2	9.3	10.3	11.4	12.3	13.2

D_1 and D_2 are as previously defined and

$$H = v_{12} D_2 + \frac{G_{12} h^3}{6} = v_{21} D_1 + \frac{G_{12} h^3}{6}. \qquad (9.26a)$$

This procedure is conservative for aspect ratios nearer to square and for a more accurate estimate reference may be made to Smith [1].

Where equation (9.26) is applied to a stiffener web the parameter b is equal to the depth of the web. In the case of panel buckling between hat section stiffeners the assumption that b is equal to the spacing between stiffener flanges may be unsafe because of local stiffener flexibility and conservatively b should be made equal to the spacing between stiffener centre lines.

More accurate analysis of stiffened panels will require application of FE analysis, which will also be necessary for consideration of overall buckling of panels under shear loading.

Through-thickness shear will not usually merit specific consideration. It has an effect on critical buckling loads but is generally covered in recommended safety factors. For more information see Chapter 12 of reference [8].

9.6 COMBINED LOADING

9.6.1 In-Plane Loading and Shear

This is one of the most common forms of load combination and may be allowed for using the interaction formula

$$\frac{F}{F_{cr}} + \left(\frac{Q}{Q_{cr}}\right)^2 = 1 \qquad (9.27)$$

where F and Q are the applied in-plane and shear forces respectively.

For combined shear and bi-axial compression on an orthotropic strip reference should be made to Davidson et al. [9], who gives solutions for isotropic material which may be used as a first approximation.

9.6.2 In-Plane and Lateral Loading

This combination of loads will be most significant in the decks and bottom of a ship. It is usually the case, except for minor internal decks, that the in-plane load dominates, and the procedures given in section 9.4 should be used for initial design, but with increased margins to allow for the lateral pressure. Margins must be based on circumstances, but it must be remembered that the total stress at a surface, to avoid FPF, should be less than about 25 per cent of the failure stress under worst design loads, see section 9.3.2.

In the case of an internal deck where lateral pressure predominates, then the procedures of section 9.3 should be used for initial sizing, but again increasing margins to take account of the in-plane component of load.

In either case when initial design is complete the structure must be analysed using FE or folded plate methods to calculate both design stresses and critical buckling stresses. The critical load F_{cr} is calculated in the absence of the lateral load, but for flexure the outer fibre stress must be enhanced by the interaction formula

$$\sigma' = \sigma_B \left/ \left(1 - \frac{F}{F_{cr}}\right) \right. + \sigma_A \qquad (9.28)$$

where σ_B is the stress due to lateral load alone, σ_A is the stress due to axial load alone, and σ' is the stress to be compared with the allowable fraction of yield stress (all stresses assumed compressive).

9.6.3 Other Load Combinations

For any other combination of loads there is insufficient data available for them all to be used during initial design. It will be necessary to design for the most significant load with subjective allowances for other loads and to analyse the structure subsequently using probably

FE in conjunction with the correct loads and boundary conditions.

9.7 REFERENCES

1] Smith, C.S., "Design of Marine Structures in Composite Materials", Elsevier Applied Science, London, 1990.

2] Johnson, A.F., Sims, G.D., "Design Procedures for Plastics Panels", NPL Design Guide, 1987.

3] Roark, R.J., "Formulas for Stress and Strain", McGraw Hill, New York. (latest edition)

4] Johnson, A.F., "Design Analysis for GRP Plates in Flexure", Proc. 13th Reinf. Pl. Cong., BPF, 1982.

5] Chalmers, D.W., "The Properties and uses of Marine Structural Materials", J. Mar. Struct., 1, 1988. pp 47-70.

6] Olsson, K-A., Reichard, R.P., (ed.), "Sandwich Constructions 1", EMAS, UK, 1989.

7] Allen, H.G., "Analysis of Structural Sandwich Panels", Pergamon Press, Oxford, 1969.

8] Herakovich, C.T., Tarnopolski, Y.M., (ed.), "Handbook of Composites", 2, *Structures and Design*, Elsevier, Amsterdam, 1989.

9] Davidson, P.C., Chapman, J.C., Smith, C.S., Dowling, P.J., "The Design of Plate Panels Subject to In-plane Shear and Biaxial Compression", Trans. RINA, 132, 1990. pp 267-268.

10 FINITE ELEMENT ANALYSIS OF COMPOSITES

10.1 INTRODUCTION

During the last two decades, the finite element (FE) method of analysis has rapidly become a very common technique for the solution of complex problems in engineering by computer codes. One of the oldest fields of application of the FE method, and the most interesting in this instance is the structural one.

The objective of this Chapter is to provide a broad overview of the FE techniques, in general for structural application, and in particular to illustrate the possibility of applying them to composite materials. As a first step, it is necessary to provide a brief resume of the physical and mathematical bases which represent the fundamental background for the FE method.

The second step aims to give an understanding of the basic features of the method itself and an idea of the typical elements. Thirdly, the introduction of higher order/complex elements is the best way to explain the extension of this technique to layered shells and thus to composite materials.

In this context, and in particular as concerns the theoretical aspects, in order to simplify the treatment, we will refer to mechanical loads and to static analysis, but in principle all the concepts can be extended to thermal loads and dynamic analysis.

10.2 BACKGROUND TO FE TECHNIQUES

10.2.1 Fundamentals of Elasticity

For the determination of the distribution of displacement and stress in a structure under external loading, it is necessary to obtain a solution to the basic equations of the theory of elasticity and satisfy the imposed boundary conditions on forces and/or displacements. For a general three-dimensional structure, these equations are:

- six strain-displacement equations

- six stress-strain equations
- three equations of equilibrium

Thus there are fifteen equations to obtain a solution for fifteen unknown variables, three displacements, six strains and six stresses. These equations reduce to eight for two-dimensional problems.

In the following sections, the basic equations of the theory of elasticity are reported.

10.2.1.1 Strain-displacement equations

Let the three displacements describing the deformed shape of a structure be

$$
\begin{aligned}
u_x &= u_x(x,y,z) \\
u_y &= u_y(x,y,z) \\
u_z &= u_z(x,y,z)
\end{aligned}
\tag{10.1}
$$

The strain corresponding to the deformed structure can be expressed by the following relations which, for small deformations, are linear [1,2]:

$$
\begin{aligned}
\varepsilon_{xx} &= \frac{\partial u_x}{\partial x} \\
\varepsilon_{yy} &= \frac{\partial u_y}{\partial y} \\
\varepsilon_{zz} &= \frac{\partial u_z}{\partial z}
\end{aligned}
\tag{10.2}
$$

for the normal strains, and

$$
\begin{aligned}
\varepsilon_{xy} &= \varepsilon_{yx} = \frac{\partial u_y}{\partial x} + \frac{\partial u_x}{\partial y} \\
\varepsilon_{yz} &= \varepsilon_{zy} = \frac{\partial u_z}{\partial y} + \frac{\partial u_y}{\partial z} \\
\varepsilon_{zx} &= \varepsilon_{xz} = \frac{\partial u_x}{\partial z} + \frac{\partial u_z}{\partial x}
\end{aligned}
\tag{10.3}
$$

for the shear strains.

Using the symmetry relationship

$$
\varepsilon_{ij} = \varepsilon_{ji} \qquad i,j = x,y,z
\tag{10.4}
$$

between all the shear strains, the strain state in three-dimensions is thus described by only six strain components.

10.2.1.2 Stress-strain equations

The stress-strain relationship for a general anisotropic elastic material is given by the generalised form of Hooke's law:

$$\sigma_i = C_{ij}\,\varepsilon_j \qquad i,j = 1,...,6 \tag{10.5}$$

with σ_i and ε_j respectively the stress and strain components in contracted notation and C_{ij} the stiffness matrix (with $C_{ij} = C_{ji}$); σ and ε the equivalent tensor notations for the stresses and strains is as below:

$$
\begin{aligned}
\sigma_1 &\rightarrow \sigma_{xx} & \varepsilon_1 &\rightarrow \varepsilon_{xx} \\
\sigma_2 &\rightarrow \sigma_{yy} & \varepsilon_2 &\rightarrow \varepsilon_{yy} \\
\sigma_3 &\rightarrow \sigma_{zz} & \varepsilon_3 &\rightarrow \varepsilon_{zz} \\
\sigma_4 &\rightarrow \sigma_{yz} = \tau_{yz} & \varepsilon_4 &\rightarrow \gamma_{yz} = 2\,\varepsilon_{yz} \\
\sigma_5 &\rightarrow \sigma_{zx} = \tau_{zx} & \varepsilon_5 &\rightarrow \gamma_{zx} = 2\,\varepsilon_{zx} \\
\sigma_6 &\rightarrow \sigma_{xy} = \tau_{xy} & \varepsilon_6 &\rightarrow \gamma_{xy} = 2\,\varepsilon_{xy}
\end{aligned} \tag{10.6}
$$

Because of $C_{ij} = C_{ji}$, the stiffness matrix has only 21 independent constants and, in the particular case of orthotropy, which is the most common situation for engineering applications of composite materials, the stiffness matrix has only nine independent constants. It reduces to [4]:

$$
C_{ij} =
\begin{bmatrix}
C_{11} & C_{12} & C_{13} & 0 & 0 & 0 \\
C_{12} & C_{22} & C_{23} & 0 & 0 & 0 \\
C_{13} & C_{23} & C_{33} & 0 & 0 & 0 \\
0 & 0 & 0 & C_{44} & 0 & 0 \\
0 & 0 & 0 & 0 & C_{55} & 0 \\
0 & 0 & 0 & 0 & 0 & C_{66}
\end{bmatrix} \tag{10.7}
$$

By inverting the stiffness matrix, equation (10.5) can be written as:

$$\varepsilon_j = a_{ij}\,\sigma_i \qquad i,j = 1,...,6 \qquad\qquad (10.8)$$

where a_{ij} is the compliance matrix.

In this case (of an orthotropic material) the stiffness and compliance matrices can be written in terms of the engineering constants E_1, E_2, E_3 (Young's moduli in 1, 2 and 3 directions), ν_{ij} (Poisson's ratios) and G_{23}, G_{31}, G_{12} (shear moduli in the 2-3, 3-1 and 1-2 planes).

The compliance matrix, in terms of the above mentioned constants, is written below:

$$[a_{ij}] = \begin{bmatrix} \dfrac{1}{E_1} & -\dfrac{\nu_{21}}{E_2} & -\dfrac{\nu_{31}}{E_3} & 0 & 0 & 0 \\[2mm] -\dfrac{\nu_{12}}{E_1} & \dfrac{1}{E_2} & -\dfrac{\nu_{32}}{E_3} & 0 & 0 & 0 \\[2mm] -\dfrac{\nu_{13}}{E_1} & -\dfrac{\nu_{23}}{E_2} & \dfrac{1}{E_3} & 0 & 0 & 0 \\[2mm] 0 & 0 & 0 & \dfrac{1}{G_{23}} & 0 & 0 \\[2mm] 0 & 0 & 0 & 0 & \dfrac{1}{G_{31}} & 0 \\[2mm] 0 & 0 & 0 & 0 & 0 & \dfrac{1}{G_{12}} \end{bmatrix}. \qquad (10.9)$$

The equations (10.5) and (10.8) are also valid in two dimensions. The 2-D representation (and in particular the plane stress formulation) is useful to understand the behaviour of a composite laminate in its lamination plane.

10.2.2 Application of Energy Theorems
In this section the following quantities, reported with their definitions and mathematical meanings, will be used:

Symbol	Definition
$W = \int F\,du$	Work

$$W^* = \int u \, dF$$ Complementary work

$$\overline{U}_i = \int \sigma \, d\varepsilon$$ Density of strain-energy

$$\overline{U}_d = \int \sigma \, d\varepsilon$$ Density of strain-energy of total deformation

$$\overline{U}_i^* = \int \varepsilon \, d\sigma$$ Density of complementary strain-energy

$$\overline{U}_d^* = \int \varepsilon \, d\sigma$$ Density of complementary strain-energy of total deformation

$$U_i = \int \overline{U}_i \, dV$$ Strain-Energy

$$U_d = \int \overline{U}_d \, dV$$ Strain-energy of total deformation

$$U_i^* = \int \overline{U}_i^* \, dV$$ Complementary strain-energy

$$U_d^* = \int \overline{U}_d^* \, dV$$ Complementary strain-energy of total deformation

$$U_e$$ Potential of external forces, where the following relationship exists:
$$\delta U_e = -\delta W$$

$$U_e^*$$ Complementary potential of external forces, where the following relationship exists: $\delta U_e^* = -\delta W^*$

$$U = U_i + U_e$$ Potential energy

$$U^* = U_i^* + U_e^*$$ Complementary potential energy

As the differential equations of elasticity for complex structures can be solved in closed form only in a very few special cases, it is necessary to refer to the concept of strain and potential energy (and to the relevant theorems), in order to solve them for actual applications.

It can be demonstrated that all the energy theorems can be derived from two energy principles:

- The principle of virtual work (or virtual displacements)   ⇒ $\delta U_i = \delta W$
- The principle of complementary virtual work (or virtual forces)   ⇒ $\delta U_d^* = \delta W^*$

These two principles are the basis of any strain-energy approach to structural analysis. A brief summary of the energy theorems, according to [2], is given below.

Theorems based on the principle of virtual work

a. The principle of a stationary value of total potential energy (principle of minimum potential energy):

$$\delta_\varepsilon U = \delta(U_i + U_e) = 0 \tag{10.10}$$

where the subscript ε means that only elastic strains and displacements are to be varied.

b. Castigliano's theorem (part I):

$$\left(\frac{\delta U_i}{\delta u_r}\right)_{T=\text{const.}} = F_r \tag{10.11}$$

where δu_r is the virtual displacement in the direction of the force F_r while keeping temperatures constant.

c. The theorem of minimum strain-energy:

$$\delta_\varepsilon U_i = 0 \tag{10.12}$$

obtainable in a strained structure only if the virtual displacements selected are equal to zero at the points of application of the forces, that is to say $\delta W = 0$.

d. The unit-displacement theorem:

$$F_r = \int_v \sigma^T \varepsilon_r \, dV \tag{10.13}$$

where σ^T is the stress matrix and δ_r is a compatible strain distribution due to a unit displacement applied in the direction of F_r.

Theorems based on the principle of complementary virtual work

a. The principle of a stationary value of total complementary potential energy (principle of minimum complementary potential energy):

$$\delta_\sigma U^* = \delta(U_d^* + U_e^*) = 0. \tag{10.14}$$

b. Castigliano's theorem (part II):

$$\left(\frac{\delta U_d^*}{\delta F_r}\right)_{T=\text{const.}} = u_r. \tag{10.15}$$

c. The theorem of minimum complementary strain-energy of total deformation (applicable only to redundant structures):

$$\delta_\sigma U_d^* = 0. \tag{10.16}$$

d. The unit-load theorem:

$$u_r = \int_v \varepsilon^T \bar{\sigma} \, dV. \tag{10.17}$$

In elastic situations the total potential energy is equal to the complementary total potential energy, but in applying the principle using theorems derived from the latter, it is necessary to specify the energy in terms of a prescribed field of stress rather than a prescribed set of displacements.

In any case the problem to solve is the minimisation of the total potential energy (or the complementary total potential energy) with respect to the nodal displacement (or stresses) which have to be treated as variable parameters.

The application to general finite elements (which cannot be solved in closed form) of this procedure is only possible by a numerical process that results in approximate solutions to problems in continuum mechanics.

The variational formulation of structural mechanics problems starting from the principle of virtual work (as mentioned, a similar procedure can also be followed starting from the principle of complementary virtual work) leads to equation (10.10) (or equation (10.14)), which represents the equilibrium equation for an elastic continuum. However, the virtual work principle can be used in much more general cases.

In mathematical physics, the quantity U (potential energy) is called the functional of the problem considered. This is not the appropriate context in which to treat these problems in depth, but, if readers would like to study this subject in more detail (and in particular in the solution of variational problems by using other methods), references [2,5] and [6] should be consulted.

10.3 THE STIFFNESS METHOD

The FE method can be used in a variety of different ways. Probably the most important formulation, and the first one to be applied, is the displacement-based method, which is simple and has very good numerical properties. The displacement-based finite element method (or FEM) can be seen as an extension of the displacement method of analysis which is used in the analysis of beam and truss structures (see, for example [2] and [7]).

The steps in a typical solution for a linear elastic structural problem using FEM can be summarised as in the following.

a. Idealisation of the structural problem, i.e., of the mathematical model for:

   • the materials
   • the geometry of the structure
   • the global behaviour of the structure

   This is a basic step required for every engineering problem apart from this specific item.

b. Subdivision of the continuum in finite elements, that corresponds to the definition of the mesh. This step is particularly important from the point of view of the FEM user, because an incorrect subdivision of the structure can cause wrong results.

c. Modelling of the displacement field for each finite element taking into account the convergence requirements; that is the definition of appropriate *functions* able to represent the displacement at any location in the element as a function of the element's nodal point displacement.

At this stage, it is important to define the difference between linear elements (rod and beam) and general finite elements. The FEM applied to a structure composed of beams and rods only, yields the exact (within the assumptions of beam theory) displacements and stresses, provided the stiffness matrices have been computed exactly (i.e. in a closed form, from the solution of the appropriate differential equations). On the other hand, for the application to a structure composed of general finite elements (plate, shells or three-dimensional elements), it is necessary to define, for each element, a function to express the displacement at any point in terms of the nodal point displacements of the structural idealisation. Usually these functions are polynomials whose unknown coefficients are treated as generalised coordinates. Because these coordinates are unknown, they need a relationship linking them to the nodal point displacements.

d. Generation of the stiffness matrices and the nodal equivalent loads for each element. In the next section the stiffness matrices for some elements will be shown. It is important to note that, for an element, the nodal displacement vector u can be related to the generalised coordinates vector α by the expression:

$$u = A\,\alpha. \tag{10.18}$$

where A is the matrix relating the two vectors. In order to obtain the strain and stress vectors, it is the essential for the inverse of A matrix to exist.

e. Assembling of the element stiffness matrices into a global stiffness matrix, taking into account the connections between the elements. The conceptual operation to be performed is "sum" the stiffness matrices of all the elements to form a global stiffness matrix to be used into the equations of the assembled structural system, this is   represented by the expression:

$$F = K U. \tag{10.19}$$

where:

K is the global stiffness matrix;
U is the vector of the displacements (and rotations) that represents the degrees of freedom of the system without the boundary conditions;
F is the vector of the nodal loads (in the sense of the virtual displacement theorem)

f. Specifying the displacement boundary conditions. By this operation the degrees of freedom corresponding to the boundary conditions have to be eliminated and a new stiffness matrix results. This matrix has to be used to solve the equations (10.18).

g. Solution of the equilibrium equations (10.18), to obtain all element nodal point displacements. A numerical procedure is adopted, with particular care given to problems of convergence.

h. Evaluation of the element stresses by equations relating displacements to strains and stresses. This is a backward procedure to be applied at each finite element.

With this step the procedure is completed and the structural behaviour of the system is known.

10.4 MATRICES FOR TYPICAL ELEMENTS
The structure has to be idealised into an assembly of structural elements. So the necessary step to be performed is the determination of the stiffness characteristics of each of these elements, in order to allow the definition of the stiffness matrix of the entire structure, as shown in the previous Chapter.

Different methods are available for the calculation of the force-displacement relationships and consequently the stiffness characteristics of the structural elements. The following methods can be used:

- Unit-displacement theorem
- Castigliano's theorem (part I)
- Solution of differential equations for the element displacements
- Inversion of the displacement-force relationships.

In this Chapter the stiffness properties of some basic elements (axial rod, beam, plane triangle) are reported. More complex plane elements will be discussed in section 5, while 3-D elements (brick) will not be treated in this context because their characteristics can be evaluated in the same way as for the above mentioned ones. In any case the application of 3-D finite elements to composite materials for marine applications are limited to few special cases.

10.4.1 Axial Rod Element

This is the simplest structural element. It has no bending stiffness and has only a one-dimensional stress distribution, see figure 10.1. For the application of the displacement method it requires two element forces (F_1 and F_2) and the corresponding displacements (u_1 and u_2).

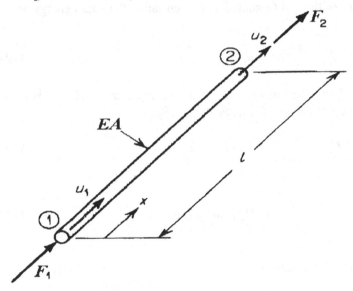

Figure 10.1. Axial rod element.

In principle the solution of the differential equations for the displacement could be used, because this element is solvable in a closed form. However, in this simple case, as an example, all the steps necessary to derive the relevant stiffness matrix will be shown, by using the Castigliano's theorem in a local coordinate system.

Using the notation of figure 10.1 the following expressions can be written:

$$\varepsilon_x = \frac{1}{l} \begin{bmatrix} -1 & 1 \end{bmatrix} \begin{bmatrix} u_2 \\ u_1 \end{bmatrix} = \frac{1}{l} (-u_1 + u_2) \tag{10.20}$$

$$\sigma_x = \frac{E}{l} \begin{bmatrix} -1 & 1 \end{bmatrix} \begin{bmatrix} u_1 \\ u_2 \end{bmatrix} = \frac{E}{l} (-u_1 + u_2) \tag{10.21}$$

where E is the elastic modulus of the material in the x direction.

Therefore, from the definition of the strain-energy and of the density of strain-energy, it follows that:

$$U_i = \frac{1}{2} \int_v \sigma^T \varepsilon \, dV = \frac{1}{2} \int_v \frac{E}{l^2} (u_2 - u_1)^2 \, dV. \tag{10.22}$$

Solving the integral on the volume $V=Al$, where A is the cross-section of the rod (assumed to be constant), the strain-energy is given by:

$$U_i = \frac{AE}{2l} (u_2 - u_1)^2. \tag{10.23}$$

Applying the Castigliano's theorem, equation (10.11), it is possible to obtain the element forces F_1 and F_2:

$$F_1 = \frac{\partial U_i}{\partial u_1} = \frac{AE}{l} (u_1 - u_2) \tag{10.24}$$

$$F_2 = \frac{\partial U_i}{\partial u_2} = \frac{AE}{l} (-u_1 + u_2) \tag{10.25}$$

and, in matrix form:

$$\begin{bmatrix} F_1 \\ F_2 \end{bmatrix} = \frac{AE}{l} \begin{bmatrix} 1 & -1 \\ -1 & 1 \end{bmatrix} \begin{bmatrix} u_1 \\ u_2 \end{bmatrix} \tag{10.26}$$

where:

$$K = \frac{AE}{l} \begin{bmatrix} 1 & -1 \\ -1 & 1 \end{bmatrix}$$ (10.27)

is the stiffness matrix of the element.

10.4.2 Beam

The beam element is assumed to be a straight bar of uniform cross-section capable of resisting axial forces, bending moments about two principal axes in the plane of its cross-section and twisting moment about its centroidal longitudinal axis, see figure 10.2.

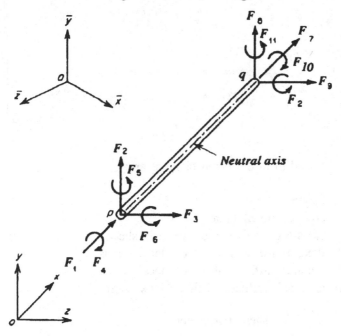

Figure 10.2. A three-dimensional beam element.

The forces acting on the beam at its extremes are: two axial forces (F_1 and F_7), four shear forces (F_2, F_3, F_8 and F_9), four bending moments (F_5, F_6, F_{11} and F_{12}), and two torsional moments (F_4 and F_{10}): the displacements u have corresponding subscripts.

The beam element is solvable in closed form and the solutions of differential equations for the element can be used. After the differentiation of the equations, the resulting complete forces-displacements relationship can be obtained in the matrix form as shown below.

$$
\begin{Bmatrix} F_1 \\ F_2 \\ F_3 \\ F_4 \\ F_5 \\ F_6 \\ F_7 \\ F_8 \\ F_9 \\ F_{10} \\ F_{11} \\ F_{12} \end{Bmatrix} =
\begin{bmatrix}
\dfrac{EA}{l} & & & & & & & & & & & \\[6pt]
0 & \dfrac{12EI_z}{l^3(1+T_y)} & & & & & & \text{Symm.} & & & & \\[6pt]
0 & 0 & \dfrac{12EI_y}{l^3(1+T_z)} & & & & & & & & & \\[6pt]
0 & 0 & 0 & \dfrac{GJ}{l} & & & & & & & & \\[6pt]
0 & 0 & \dfrac{-6EI_y}{l^2(1+T_z)} & 0 & \dfrac{(4+T_z)EI_y}{l(1+T_z)} & & & & & & & \\[6pt]
0 & \dfrac{6EI_z}{l^2(1+T_y)} & 0 & 0 & 0 & \dfrac{(4+T_y)EI_z}{l(1+T_y)} & & & & & & \\[6pt]
\dfrac{-EA}{l} & 0 & 0 & 0 & 0 & 0 & \dfrac{EA}{l} & & & & & \\[6pt]
0 & \dfrac{-12EI_z}{l^3(1+T_y)} & 0 & 0 & 0 & \dfrac{-6EI_z}{l^2(1+T_y)} & 0 & \dfrac{12EI_z}{l^3(1+T_y)} & & & & \\[6pt]
0 & 0 & \dfrac{-12EI_y}{l^3(1+T_z)} & 0 & \dfrac{6EI_y}{l^2(1+T_z)} & 0 & 0 & 0 & \dfrac{12EI_y}{l^3(1+T_z)} & & & \\[6pt]
0 & 0 & 0 & \dfrac{-GJ}{l} & 0 & 0 & 0 & 0 & 0 & \dfrac{GJ}{l} & & \\[6pt]
0 & 0 & \dfrac{-6EI_y}{l^2(1+T_z)} & 0 & \dfrac{(2-T_z)EI_y}{l(1+T_z)} & 0 & 0 & 0 & \dfrac{6EI_y}{l^2(1+T_z)} & 0 & \dfrac{(4+T_z)EI_y}{l(1+T_z)} & \\[6pt]
0 & \dfrac{6EI_z}{l^2(1+T_y)} & 0 & 0 & 0 & \dfrac{(2-T_y)EI_z}{l(1+T_y)} & 0 & \dfrac{-6EI_z}{l^2(1+T_y)} & 0 & 0 & 0 & \dfrac{(4+T_y)EI_z}{l(1+T_y)}
\end{bmatrix}
\begin{Bmatrix} u_1 \\ u_2 \\ u_3 \\ u_4 \\ u_5 \\ u_6 \\ u_7 \\ u_8 \\ u_9 \\ u_{10} \\ u_{11} \\ u_{12} \end{Bmatrix}
$$

(10.28)

Symbols referred to in the beam element are:

l length

A cross-sectional area

A_s cross-sectional area effective in shear

EI_y flexural stiffness about the local y axis

EI_z flexural stiffness about the local z axis

GJ torsional stiffness of the cross-section

$$T = \frac{12\,EI}{GA_s\,l^2} \quad \text{shear parameter}$$

10.4.3 Plane Elements

10.4.3.1 Plane stress and plane strain distribution

Plane stress is a two-dimensional stress distribution able to represent the stress-strain behaviour of thin flat plates loaded in the plane of the plate. Plane strain, on the other hand, is a two-dimensional stress distribution able to represent the behaviour of elongated bodies of constant cross-section subjected to uniform loading.

Plane stress distribution is based on the following hypothesis:

$$\sigma_{zz} = \sigma_{zx} = \sigma_{yz} = 0. \tag{10.29}$$

The three-dimensional stress-strain law represented by equations (10.5) and (10.8) reduces to a two-dimensional one in which the stress vector is, $\{\sigma\} = \{\sigma_{xx} \ \sigma_{yy} \ \sigma_{xy}\}$, and the compliance matrix is:

$$\{a_{ij}\} = \begin{bmatrix} \dfrac{1}{E_1} & -\dfrac{v_{21}}{E_2} & 0 \\[2ex] -\dfrac{v_{12}}{E_1} & \dfrac{1}{E_2} & 0 \\[2ex] 0 & 0 & \dfrac{1}{G_{12}} \end{bmatrix}. \tag{10.30}$$

Plane strain distribution is based on the assumptions that:

$$\varepsilon_{zz} = \varepsilon_{zx} = \varepsilon_{yz} = 0. \tag{10.31}$$

In this case σ_{z} is not zero, but is a linear combination of σ_{xx} and σ_{yy}. Here also the three-dimensional stress-strain law represented by equations (10.5) and (10.8) reduces to a two-dimensional one in which the stress vector is $\{\sigma\} = \{\sigma_{xx} \ \sigma_{yy} \ \sigma_{xy}\}$, the strain vector is $\{\varepsilon\} = \{\varepsilon_{xx} \ \varepsilon_{yy} \ \varepsilon_{xy}\}$, and the compliance matrix is:

$$\{a_{ij}\} = \begin{bmatrix} \dfrac{1 - v_{13} v_{31}}{E_1} & -\dfrac{v_{21} + v_{31} v_{23}}{E_2} & 0 \\[2ex] -\dfrac{v_{12} + v_{13} v_{32}}{E_1} & \dfrac{1 + v_{32} v_{23}}{E_2} & 0 \\[2ex] 0 & 0 & \dfrac{1}{G_{12}} \end{bmatrix}. \tag{10.32}$$

As can easily be seen, the greater part of marine structures can be effectively modelled by using plane stress elements, coupled with rods or beams; only in very few cases would it be useful to adopt plane strain elements.

10.4.3.2 Plane stress triangle

This is a flat two-dimensional element with a constant thickness t. For small deflection the in-plane and bending deformations are uncoupled, so the elastic properties can be evaluated separately for the in-plane and out-of-plane forces. The plane stress triangle has two degrees of freedom per node (at each of the vertices), see figure 10.3.

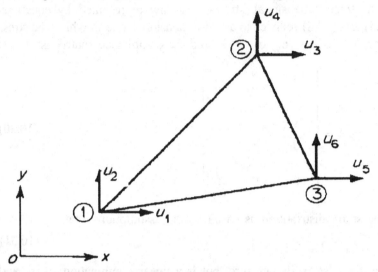

Figure 10.3. Plane stress triangle.

Because no exact stiffness relationships can be derived for this element, it is necessary to assume a displacement variation. Let this be linear in nature depending only on the displacements of the two vertices related to that edge.

If A is the area of the triangle, u_i (1,...,6) the displacements and F_i ($i = 1,...,6$) the forces at each node with the following correspondence, see figure 10.3.

$$
\begin{aligned}
u_x &= u_1 & F_x &= F_1 \\
u_y &= u_2 & F_x &= F_2 & \text{at} \quad (x_1, y_1)
\end{aligned}
$$

$$
\begin{aligned}
u_x &= u_3 & F_x &= F_3 \\
u_y &= u_4 & F_x &= F_4 & \text{at} \quad (x_2, y_2)
\end{aligned}
\tag{10.33}
$$

$$
\begin{aligned}
u_x &= u_5 & F_x &= F_5 \\
u_y &= u_6 & F_x &= F_6 & \text{at} \quad (x_3, y_3)
\end{aligned}
$$

and:

$$
\begin{aligned}
x_{ij} &= x_i - x_j \\
y_{ij} &= y_i - y_j & (i,j = 1,...3).
\end{aligned}
\tag{10.34}
$$

The force-displacement relationship is:

$$\begin{bmatrix} F_1 \\ F_2 \\ F_3 \\ F_4 \\ F_5 \\ F_6 \end{bmatrix} = K \begin{bmatrix} u_1 \\ u_2 \\ u_3 \\ u_4 \\ u_5 \\ u_6 \end{bmatrix}. \tag{10.35}$$

The stiffness matrix K can be written as the sum of two contributions:

K_n = Stiffness matrix due to normal stresses
K_s = Stiffness matrix due to shear stresses
$K = K_n + K_s$ (10.36)

$$K_N = \frac{Et}{4A(1-\nu^2)} \begin{bmatrix} y_{32}^2 & & & & \text{Symm.} & \\ -\nu y_{32} x_{32} & x_{32}^2 & & & & \\ -y_{32} y_{31} & \nu x_{32} y_{31} & y_{31}^2 & & & \\ \nu y_{32} x_{31} & -x_{32} x_{31} & -\nu y_{31} x_{31} & x_{31}^2 & & \\ y_{32} y_{21} & -\nu x_{32} y_{21} & -y_{31} y_{21} & \nu x_{31} y_{21} & y_{21}^2 & \\ -\nu y_{32} x_{21} & x_{32} x_{21} & \nu y_{31} x_{21} & -x_{31} x_{21} & -\nu y_{21} x_{21} & x_{21}^2 \end{bmatrix}.$$
(10.37)

$$K_s = \frac{Et}{8A(1+\nu)} \begin{bmatrix} x_{32}^2 & & & & \text{Symm.} & \\ -x_{32} y_{32} & y_{32}^2 & & & & \\ -x_{32} x_{31} & y_{32} x_{31} & x_{31}^2 & & & \\ x_{32} y_{31} & -y_{32} y_{31} & -x_{31} y_{31} & y_{31}^2 & & \\ x_{32} x_{21} & -y_{32} x_{21} & -x_{31} x_{21} & y_{31} x_{21} & x_{21}^2 & \\ -x_{32} y_{21} & y_{32} y_{21} & x_{31} y_{21} & -y_{31} y_{21} & -x_{21} y_{21} & y_{21}^2 \end{bmatrix}.$$
(10.38)

10.5 HIGHER ORDER/COMPLEX ELEMENTS
The elements discussed in the previous sections are characterised by

assumed linear displacements. Higher order and/or more complex elements are often used in structural analysis, and in particular they play an important role for composite materials. In the following sections, an overview of these elements, with particular reference to isoparametric plane elements and layered shells, will be presented.

10.5.1 Isoparametric Formulation

In order to understand the isoparametric formulation, it is useful to define the meaning of "natural" coordinate system. A three-dimensional natural coordinate system is one with variables η, ξ, and ζ varying from +1 to -1. The basis of isoparametric finite element formulation is the interpolation of the element coordinates and element displacements using the same interpolation functions, which are defined in a natural coordinate system.

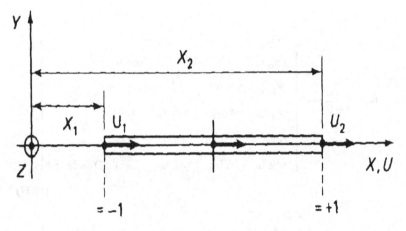

Figure 10.4. One-dimensional element for isoparametric formulation.

For example, consider a one-dimensional element, see figure 10.4, lying in the global X coordinate axis. The first step is to establish a correlation between the global coordinate X and the natural coordinate system with η, where $-1 \leq \eta \leq +1$. This correlation able to transform one system into the other is:

$$X = \frac{1}{2}(1 - \eta)X_1 + \frac{1}{2}(1 + \eta)X_2, \qquad (10.39)$$

that is to say, if $\theta_1 = 1/2(1 - \eta)$ and $\theta_2 = 1/2(1 + \eta)$:

$$X = \sum_{i=1}^{2} \theta_i X_i \qquad (10.40)$$

θ_i are the so called interpolation functions. Starting from these functions, it is easy to find the stiffness matrix for the rod (equation (10.27)).

For a general three-dimensional element, the coordinate interpolations are:

$$x = \sum_{i=1}^{q} \theta_i x_i$$

$$y = \sum_{i=1}^{q} \theta_i y_i \qquad (10.41)$$

$$z = \sum_{i=1}^{q} \theta_i z_i$$

where x,y and z are the coordinates at any point of the element and x_i, y_i and z_i $(i=1,...,q)$ are the coordinates of the q element nodes.

In this way, by isoparametric formulation, a relationship is achieved between the element displacements at any point and its nodal point displacements directly through the use of interpolation functions (also called shape functions).

This means that the transformation matrix A of equation (10.18), when global coordinates are used, is not evaluated, while the element matrices corresponding to the required degrees of freedom are obtained directly. In fact, in the isoparametric formulation, the element displacements are interpolated in the same way as the geometry, according to the expressions:

$$u = \sum_{i=1}^{q} \theta_i u_i$$

$$v = \sum_{i=1}^{q} \theta_i v_i \qquad (10.42)$$

$$w = \sum_{i=1}^{q} \theta_i w_i$$

where u,v and w are the local element displacements at any point of the element and u_i, v_i and w_i $(i=1,...,q)$ are the corresponding element displacement at its node.

To evaluate the stiffness matrix of an element, it is necessary to calculate the strain-displacement transformation matrix.

In practice it is necessary to relate the x, y and z derivatives to the η, ξ and ζ derivatives. This relation can be obtained by using the Jacobian operator J defined as:

$$J = \begin{bmatrix} \dfrac{\partial x}{\partial \eta} & \dfrac{\partial y}{\partial \eta} & \dfrac{\partial z}{\partial \eta} \\[2ex] \dfrac{\partial x}{\partial \xi} & \dfrac{\partial y}{\partial \xi} & \dfrac{\partial z}{\partial \xi} \\[2ex] \dfrac{\partial x}{\partial \zeta} & \dfrac{\partial y}{\partial \zeta} & \dfrac{\partial z}{\partial \zeta} \end{bmatrix} . \tag{10.43}$$

So, it is possible to write:

$$\begin{bmatrix} \dfrac{\partial}{\partial \eta} \\[2ex] \dfrac{\partial}{\partial \xi} \\[2ex] \dfrac{\partial}{\partial \zeta} \end{bmatrix} = J \begin{bmatrix} \dfrac{\partial}{\partial x} \\[2ex] \dfrac{\partial}{\partial y} \\[2ex] \dfrac{\partial}{\partial z} \end{bmatrix} , \tag{10.44}$$

or

$$\begin{bmatrix} \dfrac{\partial}{\partial x} \\[2ex] \dfrac{\partial}{\partial y} \\[2ex] \dfrac{\partial}{\partial z} \end{bmatrix} = J^{-1} \begin{bmatrix} \dfrac{\partial}{\partial \eta} \\[2ex] \dfrac{\partial}{\partial \xi} \\[2ex] \dfrac{\partial}{\partial \zeta} \end{bmatrix} . \tag{10.45}$$

Using equations (10.42) and (10.45), the strain-displacement transformation matrix B can be constructed, and the following expression written:

$$\varepsilon = B \, u. \tag{10.46}$$

The element stiffness matrix is:

$$K = \int_{v} B^{T} \, C B \, dV \tag{10.47}$$

where C is the elasticity matrix. Because

$$dV = \det J \, d\eta \, d\xi \, d\zeta \text{ and assuming}$$

$$F = B^T CB$$

$$K = \int_v F \, d\eta \, d\xi \, d\zeta \quad \text{is obtained.} \tag{10.48}$$

10.5.2 Numerical Integration

Equation (10.48) requires a numerical integration, because in a very few practical cases only, can it be integrated in an analytical way. The integral in this formula can be evaluated numerically using:

$$\int_v F(\eta, \xi, \zeta) \, d\eta \, d\xi \, d\zeta = \sum_{i,j,k} \alpha_{ijk} \, F(\eta_i, \xi_j, \zeta_k) + R_n \tag{10.49}$$

where α_{ijk} are weighting factors and R_n is an error matrix which in practice is not evaluated.

A very important and well assessed numerical integration procedure is the Gauss quadrature, where the basic assumption, referred to a one-dimensional case is:

$$\int_a^b F(\eta) \, d\eta = \alpha_1 F(\eta_1) + \alpha_2 F(\eta_2) + \dots + \alpha_n F(\eta_n) + R_n \tag{10.50}$$

where both the weights α_i and the sampling points η_i are variables. In isoparametric element calculations $a = -1$ and $b = +1$.

In practice it is necessary to determine the sampling points in the interval a to b; the values of these quantities have been published for different n in reference [8]. More details on this procedure and some examples of its practical application are reported in reference [5].

10.5.3 Plane Linear Isoparametric Element

In order to better understand the isoparametric elements, it is helpful to look to the formulation of a four nodes linear rectangular isoparametric plane element. The natural coordinates system be (η, ξ) defined on a domain Ω, called master element, to which corresponds to the actual element domain Ω_n, by means of the transformation T:

$$T \begin{cases} x = x(\eta, \xi) \\ y = y(\eta, \xi) \end{cases}. \tag{10.51}$$

According to equations (10.41):

$$x = \sum_{i=1}^{4} \theta_i(\eta, \xi) x_i$$
$$y = \sum_{i=1}^{4} \theta_i(\eta, \xi) y_i \qquad (10.52)$$

where the linear shape functions are defined as:

$$\theta_1(\eta, \xi) = \frac{1}{4}(1 - \eta)(1 - \xi)$$

$$\theta_2(\eta, \xi) = \frac{1}{4}(1 + \eta)(1 - \xi)$$

$$\theta_3(\eta, \xi) = \frac{1}{4}(1 + \eta)(1 + \xi) \qquad (10.53)$$

$$\theta_4(\eta, \xi) = \frac{1}{4}(1 - \eta)(1 + \xi).$$

In the same way, according to equations (10.42), there is, on the master element Ω:

$$u = \sum_{i=1}^{4} \theta_i(\eta, \xi) u_i$$
$$v = \sum_{i=1}^{4} \theta_i(\eta, \xi) v_i. \qquad (10.54)$$

The Jacobian operator J of the transformation T can be derived by expanding the expression (10.52); for the coordinate x (for example):

$$x = \frac{1}{4} A_x + \frac{1}{4} B_x \eta + \frac{1}{4} C_x \xi + \frac{1}{4} D_x \eta \xi \qquad (10.55)$$

where:

$$A_x = x_1 + x_2 + x_3 + x_4$$
$$B_x = -x_1 + x_2 + x_3 - x_4$$
$$C_x = -x_1 - x_2 + x_3 + x_4$$
$$D_x = x_1 - x_2 + x_3 - x_4$$

and A_y, B_y, C_y, D_y are defined in the same way

$$J = \frac{1}{4} \begin{bmatrix} B_x + D_x \xi & B_y + D_y \xi \\ C_x + D_x \eta & C_y + D_y \eta \end{bmatrix}. \qquad (10.56)$$

10.5.4 Isoparametric Plates and Shells

In the previous sections only flat elements have been treated. Another field of problems concerns the bending of elastic plates. To evaluate the stiffness matrix of plates and shells (which can be treated as distorted plates, or as an assembly of flat plates), including the effects of transverse deflections and rotations, it is necessary to assume a deflection u_z as a function of x and y. Hence u_z and derivatives of u_z up to a chosen order, are to be taken as nodal parameters. After this assumption has been made, the plate bending problem can be treated in the same manner as the plane stress problem.

The adoption of isoparametric formulation for the definition of shell elements is particularly important because the correspondence between the actual element coordinate system and the natural coordinate system allows two-dimensional elements to be mapped into two or three-dimensional surfaces.

10.5.5 Extensions to Layered shells

All the previous sections have been developed in order to understand how FEM can be used to represent the behaviour of a structure in terms of displacements, stresses and strains. The application of the method to composite materials can be summarised in two main approaches:

• the first one, which is the simplest, considers the composite laminate as a homogeneous material, analysing the effects of the constituent materials (and laminae) as averaged properties of the composite;

• the second one, starting from the properties of the single lamina (theoretically or experimentally determined), considers the laminate in its actual configuration, as a stack of laminae with various orientations of principal materials in the laminate.

As regards the first approach, it is possible to apply the FEM to composites by using plate or shell elements and adopting an orthotropic representation of the material, (see section 10.2.1). In this case some simplified assumptions about the material behaviour have to be made. In particular a lamination theory [9] able to take into account the contributions of the different laminae to the actual laminate characteristics, has to be used.

The second approach, which is more reliable, because it is able to represent the mechanical behaviour of each lamina and the interaction between the different laminae, is strictly connected to the introduction of layered shells in finite element technique.

For this kind of element, in order to model the material behaviour of composites, the orthotropic characteristics of the laminae must be represented in their own planes, and then integrated through the thickness on a layer-by-layer basis. Because of the different patterns of deformation of the laminae (mainly due to the different orientations and consequently to the different mechanical characteristics), high shear stresses (the so called interlaminar shears) occur between the laminae. These shear stresses ought to be accurately represented in the model.

Because, in general, most composite structures are thin-walled, the most natural geometric model for slender composite structures is the multi-laminate shell element. Several of these elements have been proposed by different authors, (see for example reference [10]). The most commonly used are the triangular elements with three and six nodes and the quadrilateral elements with four and eight nodes.

Consider now, the mathematical formulation of this kind of element. For example, an element of constant thickness t composed of a finite number of thin orthotropic layers of uniform thickness bonded together such that each layer is identified by an integer k (k = 1,...,N). The material properties and the thickness of each layer may be entirely different. For each layer, a two-dimensional local stress-strain relation can be written in the form of equation (10.5), as:

$$\begin{bmatrix} \sigma_{xx} \\ \sigma_{yy} \\ \sigma_{xy} \end{bmatrix} = \begin{bmatrix} C_{11} & C_{12} & C_{16} \\ C_{12} & C_{22} & C_{26} \\ C_{16} & C_{26} & C_{66} \end{bmatrix} \begin{bmatrix} \varepsilon_{xx} \\ \varepsilon_{yy} \\ \varepsilon_{xy} \end{bmatrix} . \tag{10.57}$$

If the general shear deformation plate theory [11] is used, a displacement field $\{u\}$ can be assumed, constituted by three displacement components u, v and w and the two rotation φ_x and φ_y. From these data, it is possible to evaluate the strain-energy of the plate U_i defined (per unit area) by the following expression:

$$U_i = \frac{1}{2} \sum_{k=1}^{N} \int_{z_{k-1}}^{z_k} \sigma^T \varepsilon \, dz \tag{10.58}$$

where z_k and z_{k-1} represent the coordinates of the two extremes of the kth layer.

Starting from the definition of the displacement field $\{u\}$, it is also possible to evaluate the element stiffness matrix, including the in-plane and bending characteristics. Several numerical-experimental comparisons have shown that these type of multi-layered elements can be effectively used for the structural analysis of composite shell structures.

10.6 REFERENCES

1] Timoshenco, S., Goodier, J.N., "Theory of Elasticity", 2nd edition, McGraw-Hill Book Company, New York, 1951.

2] Przemieniecki, J.S., "Theory of Matrix Structural Analysis", McGraw-Hill Book Company, New York, 1968.

3] Lekhnitskii, S.G., "Theory of Elasticity of an Anisotropic Body", MIR Publishers, Moscow, 1981. (English translation)

4] Jones, R.M., "Mechanics of Composite Materials", Hemisphere Publishing Corporation, New York, 1975.

5] Bathe, K.J., Wilson, E.L., "Numerical Methods in Finite Elements Analysis", Prentice-Hall Inc., Englewood Cliffs, New Jersey, 1976.

6] Zienkiewicz, O.C., "The Finite Element Method in Engineering Science", McGraw-Hill Book Company, London, 1971.

7] Baldacci, R., "Scienza delle Costruzioni" (Vol. II), UTET, Torino, 1976. (In Italian)

8] Loxan, A.N., Davids, N., Levenson, A., "Table of the Zeros of the Legendre Polynomials of Order 1-16 and the Weight Coefficients for Gauss' Mechanical Quadrature Formula", Bull. Am. Math. Soc., **48**, 1942.

9] Pister, K.S., Dong, S.B., "Elastic Bending of Layered Plates", J. App. Mech. Div., ASCE, October 1959.

10] Noor, A.K., Mathers, M.D., "Shear-flexible Finite-element Models of Laminated Composite Plates and Shells", NASA, TND-8044, 1975.

11] Whitney, J.M., Pagano, N.J., "Shear Deformation in Heterogeneus Anisotropic Plates", J. App. Mech., **37**, 1031, 1970.

11 THEORETICAL PREDICTIONS OF FAILURE MECHANISMS AND STRENGTH

11.1 INTRODUCTION

In this Chapter, failure mechanisms in polymeric composite materials will be reviewed with an emphasis on how to predict materials failure and strength. The Chapter covers the simple case of unidirectional (UD) composites through to laminates, notches, compression behaviour, damage effects, environmental behaviour and fatigue. Inevitably the number of theories relating to this topic are numerous and in this Chapter only the major ones have been reviewed with apologies for those omitted due to lack of space and time. Within these constraints this review will inevitably be fairly shallow, but reference to more detailed treatise will be made for the reader to follow up if desired. The bulk of the Chapter will comprise an overview of the academic approaches to composites predictive modelling.

Before presentation of specific predictive models, it is worthwhile first considering the classes of models available. Predictive models for composite strength generally fall into three categories, based either on general failure criteria, or what is loosely called damage mechanics or fracture mechanics. The first category, general failure criteria, includes maximum strain, Tsai-Wu [1] etc. as well as those based on statistical fitting of data, popular in modelling fatigue behaviour. The second category, damage mechanics, covers all those based on the identification of failure modes and subsequent attempts to model behaviour, such as those by Beaumont [2] at Cambridge University and the team at Surrey University [3]. This approach often suffers from the disadvantage that information must be acquired on how the materials fail prior to the analysis and frequently does not include the capability to predict how failure occurs. The last category is energy based, often assuming fracture mechanics. This can be quite successful when the damage developed is essentially a single crack or can be modelled as a single crack. Unfortunately, in many instances

280

in composites, complex damage zones are developed involving cracks in many directions and these do not easily lend themselves to the application of fracture mechanics.

11.2 UNIDIRECTIONAL COMPOSITES
The fundamental case of unidirectional laminae, how they fail and how to predict strength from the properties of the fibres, resin and interface will be considered first.

11.2.1 Tension
In tension the fibres must be seen as the major reinforcing elements and failure of these will constitute composite failure. In general, models seeking to predict UD lamina strength in tension from the fibre, matrix and interface properties have been probabilistic/mechanistic in nature. Models tend to be developed specifically for composites rather than attempting to modify models developed for other materials. The fibrous nature of the materials dictates this.

The one major exception is the rule-of-mixtures (ROM), which is still widely used but only really works by chance. This relates the strength of the lamina to the fibre strength, factored by the volume fraction of fibres as:-

$$S_c = S_f V_f + S_r V_r,\qquad\qquad(11.1)$$

where S is the strength, V is the volume fraction and suffix's c, f and r relate to composite, fibre and resin respectively. This works at all only because the data used for the fibre strength is often supplied by the fibre manufacturers, who are in the habit of quoting fibre strength measured by an impregnated tow test. Since this test is a composite test anyway, it is not surprising that the ROM works. The only potential problem then is that the fibre manufacturers use simple resins for their impregnated tow tests and in some cases aerospace grade resins lead to poorer translation of the fibre strength into the composite and the ROM then overestimates composite strength. This is particularly true for many of the newer high performance fibres now available. If the real fibre strength, as determined by single filament tests is used in the ROM, the lamina strength predictions are generally much poorer.

To do better than the ROM required the use of probabilistic models. Much of the early work in this field was done way back in the 1960s and 1970s by Rosen and Zweben [4] and others. It was recognised that fibres had a significant scatter in strength, largely determined by a distribution of flaws along their lengths; fibres broke with the weakest first etc. When a fibre broke, experimental evidence on polymeric composites showed that this was contained and fast transverse fracture perpendicular to the fibres, as observed in some early MMC systems, did not occur. The load was then assumed to be distributed in some way onto the neighbouring fibres. Using Weibull probability theory, equations to predict the stress when the first fibre break occurred were developed. This generally did not agree well with composite strengths but better agreement was obtained for stresses equating to the first appearance of a doublet, two adjacent fibre breaks, or a triplet. More recently Monte-Carlo simulations on modern computers have eliminated the need to make these arbitrary assumptions about how many fibre breaks are required to cause composite failure [5]. Indeed it is relatively simple to set up a large array representing fibres, in a computer program, and increment the applied stress and eventually the fibres break in an unstable manner and failure occurs. The only assumption then is how neighbouring fibres carry the load shed by a broken fibre. Agreement with experiment can be quite good.

Clearly these models make no allowance for the role of the matrix resin or the fibre/matrix interface. The importance of these factors in affecting lamina strength has been appreciated only recently and there is much work in progress at present in the UK generating understanding and predictive models in this field [6].

11.2.2 Compression

In attempting to model the UD compressive strength of composites, the first problem is how to measure experimentally what is being modelled. Indeed designers might argue that structures rarely experience true compression failures as instability usually precedes compressive failure. To pursue this matter would occupy too much space, but suffice it to note that there have been significant advances in recent years which appear to have enabled true compressive strengths of materials to be measured [7]. How relevant these are to structures is a debate for another time.

The early models of compressive failure derived from Rosen and his fellow workers in the USA [8]. They argued that compressive strength was determined by fibre microbuckling involving elastic matrix deformation. By applying a strain energy balance to this mode of failure an expression for UD compressive strength was obtained:-

$$S_c = G_m/(1-V_f) + \pi^2 E_f d^2 V_f/3L^2.$$ (11.2)

where V_f and E_f are fibre-volume fraction and modulus. G_m is the matrix shear modulus, d the fibre diameter and L the buckling wavelength. Given a composite that fails by elastic microbuckling, good agreement can be obtained with experiment. Figure 11.1 shows data from Cambridge University [9] for a spaghetti/elastomer model material which agrees well with the Rosen equation.

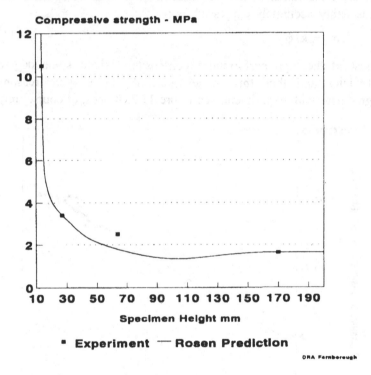

Figure 11.1. Elastic microbuckling (Rosen prediction versus experiment) spaghetti/ elastomer model system.

The Rosen expression has been found to greatly overestimate the compressive strength of the more usual polymer based composites. This is perhaps partly because it makes no allowance for plastic yield of the matrix. Budiansky [10] identified the shear yield stress k of the matrix and the initial fibre misalignment ϕ of the fibres as the main factors controlling the compressive stress when plastic microbuckling occurs. For a perfectly rigid plastic body Budiansky showed that the compressive strength is given by:-

$$\sigma_c = k^*/\phi \tag{11.3}$$

where

$$k^* = k\left(1 + (\sigma_{\tau y}/k)^2 \tan^2 \beta\right)^{1/2} \tag{11.4}$$

$\sigma_{\tau y}$ is the yield stress of the composite transverse to the fibre direction. The angle ϕ is the fibre misalignment and β is the kink-angle of the fibres in the microbuckle. For most composites this is reasonably accurately expressed as :-

$$\sigma_c = k/\phi. \tag{11.5}$$

Most of the high performance composites exhibit some degree of plasticity and thus this model generally gives quite reasonable agreement with experiment, see figure 11.2. It does, of course, rely on

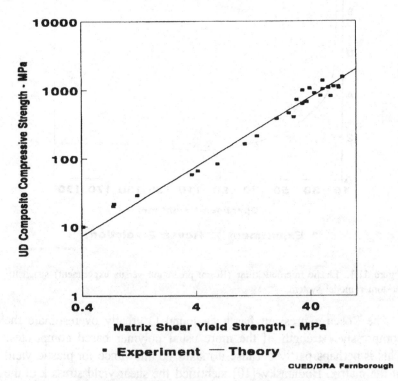

Figure 11.2. Plastic microbuckling (Budiansky prediction versus experiment).

a knowledge of the fibre misalignment, not an easy parameter to measure or indeed control.

The last major mechanism of compressive failure is by failure of the fibres either by crushing or by shearing. The latter is frequently observed in high modulus carbon fibres. If failure is by shear, then a reasonable estimate of compressive strength can be made by assuming that tensile and compressive strengths are equal. No useful models currently exist to predict the columnar crushing failure of the fibres from fibre microstructural details.

11.3 PLAIN LAMINATES
The success of predictive modelling of plain laminates hinges on developing failure criteria and basing these on observed physical processes.

11.3.1 Laminae
The predictive models developed are usually set up to deal with the individual laminae, as layers of a laminate, and how these are amalgamated into laminate predictions will be dealt with later. The major difference between dealing with laminae and the UD materials discussed above is the need to be able to describe behaviour with the stress direction at any angle to the fibres and indeed to cope with stress situations other than simple uniaxial loading situations described in the last section.

The most frequently used criteria are empirical and are extensions of those used for isotropic materials such as metals. These are the ones that come into the first category of general criteria. It is often assumed that once the strengths of a UD ply or fabric have been determined in the five main loading modes i.e.:

- Longitudinal tension and compression
- Transverse tension and compression
- Longitudinal shear

then failure criteria can be determined and strength predicted for any orthotropic material subjected to any combination of stresses. This is the essence of approaches such as those of Tsai [11] and Tsai-Wu [1] and the application of isotropic type criteria such as Hill [12] and Von Mises. The problems tend to occur in combined stress situations where load interactions may not be as predicted from these empirical models. Other approaches are attempting to introduce an element of micromechanics into the failure criteria to add some physical credence to the interactive terms in their predictive equations.

The simplest failure criteria are those based on maximum stress or strain. These simply state that, for a single ply, failure occurs when the applied stress or strain equals the failure value for that loading mode. No account is taken of load interactions, but this approach can be quite suitable when the loading is predominantly in one direction with only small loads in others. For example, the assumption that a laminate will fail when the UD plies reach their axial failure strain works well in many cases and the models used for UD composites can then be applied. Figure 11.3 shows the maximum strain criterion

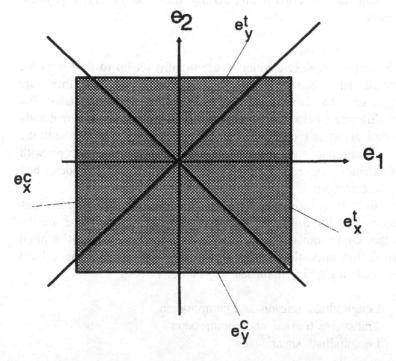

Figure 11.3. Plot of maximum strain criterion in the zero shear strain plane.

plotted in the zero shear strain plane, which shows this quite well. Clearly the lack of interactive terms make this criterion unsuitable when the loading situation is more complex than predominantly uniaxial. In addition, since Poisson's ratio is never zero, there will always be some coupling between the different strain components, so that the stress and strain based criteria will give different results except for specific cases.

Various quadratic criteria have been proposed in attempts to introduce an element of interaction between the failure modes. A typical plane stress criterion based on Tsai's adaption of the Hill criterion is as follows:

$$\left(\frac{\sigma_1}{\sigma_{1y}}\right)^2 - \frac{1}{r}\frac{\sigma_1}{\sigma_{1y}}\frac{\sigma_2}{\sigma_{2y}} + \left(\frac{\sigma_2}{\sigma_{2y}}\right) + \left(\frac{\tau_{12}}{\tau_{12y}}\right) = 1 \qquad (11.6)$$

where

$$\frac{1}{r} = \frac{\sigma_{1y}}{\sigma_{2y}}.$$

This is shown schematically in figure 11.4. No distinction is made in

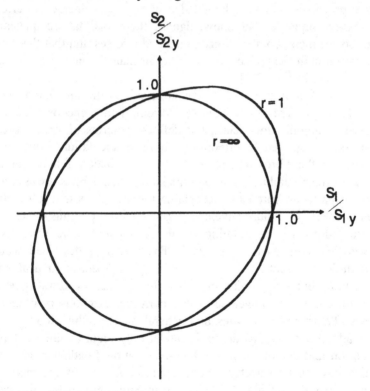

Figure 11.4. Hill yield criterion.

this case between tensile and compressive loading, although separate equations could be set up for each. σ_1 and σ_2 are the applied stresses in the longitudinal and transverse directions and the subscript y denotes their corresponding yield strengths. Since composites rarely yield it is unclear what this means, but it could be assumed that this would equate with composite failure. The term r is the ratio of the transverse and longitudinal yield stresses. This produces a failure surface as shown in figure 11.4 for zero shear stress. Again, this can be drawn in 3-dimensions, with shear stress in the third dimension. The shape depends on the term r. These criteria can equally well be translated into strain space, but any non-zero through-thickness strain is ignored, so the failure criterion is still one based on plane stress and not plane strain.

Hart-Smith would probably rebuke the failure surface type diagrams in the previous figures and indeed has cast criticism on these types of approaches. His argument is that it is meaningless to draw smooth failure envelopes through measured strengths that represent different failure processes without a knowledge of how they interact. In effect, Hart-Smith suggests that knowledge of the actual failure modes is required for many combinations of applied stresses and that this must be done through the application of micromechanics knowledge of how the materials behave [13].

The Hart-Smith failure criterion is based on the maximum shear stress failure, or Tresca criterion, for isotropic homogeneous materials, as used commonly for metals, but defined in terms of strain instead of stress. As applied to orthotropic laminae, the failure criterion is derived for the fibres and refers to fibre-dominated failure not the lamina or matrix failures. Through-thickness shear strains are assumed to be insignificant (probably acceptable for stiff fibres in a soft matrix as in CFRP). The failure envelope is generated by plotting the critical failure strains for each failure mode anticipated. The result is a skewed hexagon, as in figure 11.5. The model as described would seem little different to the other arbitrary approaches to deal with interaction, but Hart-Smith modifies this further by drawing extra lines on the failure envelope which correspond to observed failure modes. Transverse properties and general matrix failure modes have to be added this way, as the basic model only requires failure strains in tension and compression and Poisson's ratios. Loading modes and lamina have to be selected to avoid matrix failure, as the model is based essentially on fibre behaviour. Hart-Smith argues that this need not be a significant limitation, as all practical laminates have 3 or 4 fibre directions and hence failure should always be fibre dominated. Indeed, designers should feel that if a composite structure fails in a matrix dominated mode then they have not done their job properly! However, it might be reasonably assumed that delamination is a

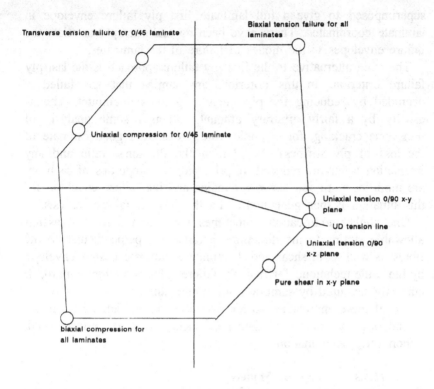

Figure 11.5. Typical failure envelope by Hart-Smith criterion.

matrix dominated failure mode that certainly is important in real structures. Again, in-plane shear can be accommodated by producing 3-D envelopes.

Perhaps the Hart-Smith approach does offer some improvements; certainly the inclusion of an element of composite mechanics must be a good thing and provide the basis for further progress.

11.3.2 Laminates
The models described so far relate to laminae or individual laminate layers. Most real composite structures will comprise many layers in laminated form, thus consideration must be given to what failure criteria may be used for these laminates.

The simplest is the first ply failure criterion. In this case the failure envelope for an on-axis ply is first produced, by whichever methods described in section 11.3.1 is favoured. Then the off-axis failure envelopes are generated from this envelope by a rigid body rotation of the envelope to the particular ply angle. The envelopes are then superimposed to give a full laminate first ply failure envelope in laminate coordinates. There have been no real attempts to relate the failure envelopes to the modes of failure of the laminates.

The main alternative to the first ply failure approach is the last ply failure criterion. In this criterion, any lamina that has failed is degraded by reducing the ply moduli by a fixed amount. This is usually by a fairly arbitrary amount, although some analysis of transverse cracking, for example, can give quite a good estimate of the loss of ply stiffness. In addition the Poisson's ratio and any interaction terms are reduced. In principle, the properties of each ply are modified until the last ply remains and the failure envelope for this situation is then generated as for the first ply failure analysis.

One slight modification sometimes used in design is to assign allowable strains to the directions parallel and perpendicular to the fibres as well as in shear and determine a laminate design envelope by the same technique for first ply failure. This is a simple approach currently favoured by some aerospace companies.

All of these approaches suffer from the serious deficiency of not considering interlaminar damage, such as edge effects and, importantly, delamination.

11.3.3 Mechanistic Models

One thing emphasised in the failure criteria described so far is the importance of plies with fibres in the principal stress direction as major load carriers. The criteria such as maximum strain really depend on this being the case to be successful at predicting strength. However, surveying the literature on micromechanical modelling of composites large numbers of papers concerned with damage development in angle plies will be found and rather less on what happens to the unidirectional plies. This may be partly because of the difficulties associated with examining damage processes in opaque UD layers, where the fibre level scale of damage is at least an order of magnitude finer than ply level damage. Nevertheless some good work has been done studying fibre scale damage processes, notably by Jameson [14] and Reifsnider [15] in the USA.

The work on damage studies at ply level is still of importance; as mentioned earlier there is a need to describe delamination, which is very difficult to include in the simple failure criteria models, but possible with the damage mechanics approach. The work by Beaumont and co-workers at Cambridge [2] and Guild, Ogin and Smith [3] are good examples of this approach. Using this method it is possible to make predictions of the extent of damage produced, perhaps transverse layer cracks, as a function of applied stress or fatigue cycles etc., for particular laminates. It can be useful in assessing the sensitivity of particular laminates to certain types of damage. The methods are based on stress analysis with the criteria for failure often based on fracture mechanics. For example, Beaumont uses two values of strain energy release rate, one for delaminations and one for splits, in his calculations. In these types of approaches there is frequently a lot of 'curve fitting' with experimentally determined parameters which probably vary with laminate type, material etc. This makes this type of technique of limited use in design, as many parameters must be determined for specific materials and laminates. However, useful design rules, such as how thick layers should be to minimise damage etc., have been developed and used successfully. The approach of trying to base predictive models on physical evidence rather than mathematical convenience, is commendable. There is, however, a long way to go before these techniques are likely to be used for predictive design purposes.

11.4 LAMINATES WITH STRESS CONCENTRATIONS

In reviewing the literature for failure criteria, it is clear that many researchers have spent a lot of effort modelling plain materials and it might asked why this is the case. Cynically it could be proposed that this is because modelling plain laminates is a less demanding task than coping with stress concentrators! If questioned on this matter, many would inevitably note that they expect to move onto the more complex situation with stress concentrators, but they often do not do so! Yet clearly real structures include a variety of features such as holes, cut-outs, etc. that will probably dominate the failure process. One conclusion from this review might be that there should be more effort devoted to the more complex design critical stress cases.

11.4.1 Notches - Tension

The earliest attempts to predict notched strength of composites were fracture mechanics based, and the work by Waddoups and associates is probably typical of this approach [16]. They applied the simple Irwin type relationship effectively with a small allowance to crack length essentially to cover the damage zone at a crack tip. This was assumed a constant for each laminate. Comparisons could be made with experiment and by varying this parameter, essentially a characteristic distance, reasonable agreement could be obtained with experiment, see figure 11.6. It should be noted, however, that the fit with the data suggests that the characteristic distance is not independent of hole size, apparently increasing with increasing hole size.

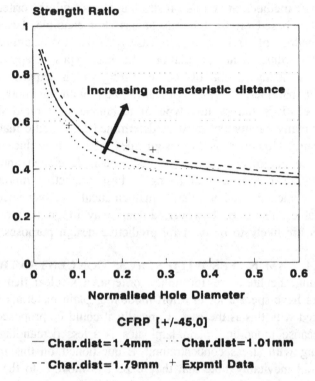

Figure 11.6. Waddoups characteristic distance model comparison with notched tensile data.

The most widely used formulation for notched strength prediction is probably that based on a modification of this characteristic distance approach and was first postulated by Whitney and Nuismer, usually referred to as the point stress and average stress criteria [17]. This approach differs from the work of Waddoups in that fracture mechanics as such is not employed. Indeed, the application of fracture mechanics to composites is questionable when single cracks of the type observed in metals rarely occur (except perhaps for delaminations).

The point/average stress criteria have proven attractive since they consider the stress state remote from the free edge and thereby do not directly address the complex damage phenomenon at the notch tip.

For this reason it is a relatively straightforward concept which is popular with designers. The point stress criterion assumes that failure occurs when the stress at some characteristic distance from the notch equals the strength of the unnotched laminate. The average stress criterion is a little different, assuming that failure occurs when the average stress over some characteristic distance equals the unnotched laminate strength. These are shown schematically in figures 11.7 and 11.8. These stress conditions can be solved for orthotropic plates to

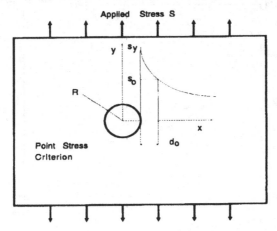

Figure 11.7. Point stress criterion.

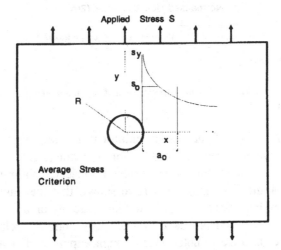

Figure 11.8. Average stress criterion.

yield equations for strength based only on elastic properties, geometry, the unnotched strength and the critical characteristic distance.

Typical comparisons with experiment are shown in figure 11.9 [18]. The results look good, but it must be remembered that these are effectively curve fits, since the characteristic distance is determined from the experimental data, so these techniques perhaps are not true predictive models.

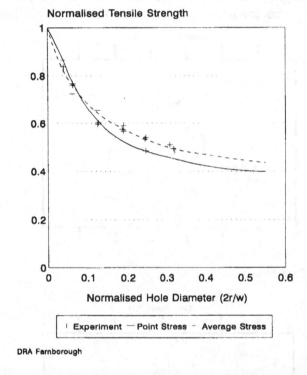

DRA Farnborough

Figure 11.9. Comparison between experiment and prediction for point and average stress criterion.

The clear failing of these semi-empirical techniques is that they require the experimental evaluation of a characteristic dimension remote from the notch boundary at which a selected failure criterion is applied. Numerous researchers have shown that this dimension is dependent on the materials system and laminate lay-up and also notch size. Some researchers have made assumptions relating the characteristic distance to notch size through exponential relationships to try and allow for this, but with limited success and they rapidly loose sight of the physical aspects of the problem. The characteristic distance is clearly not a material property, thus the models will not predict the effect of lay-up - some notch sensitivity parameter has to be measured for every lay-up. The models also offer no insight into the failure process and do not significantly reduce the amount of experimental testing required to characterise the laminate.

The problem of developing a model that has a sound physical basis was tackled by Potter [19] and this model probably represents the best available at present for predicting notched tensile strength, indeed UK aerospace companies have already made some use of this in design. The failure criterion in this model was derived from the consideration of the physical conditions necessary to initiate a chain reaction of fibre failures in the primary load-bearing plies. For failure to occur the stress at the notch tip must be sufficient to fracture the first fibre and the stress gradient at the notch tip must be low enough to permit a chain reaction of subsequent fibre failure. The key criterion for the unstable fibre failure sequence and thus crack propagation was given by:

$$\left| \frac{\partial \sigma_x}{\partial y} \right| \leq \frac{\delta D}{s}. \tag{11.7}$$

where σ_x is the laminate stress in the axial x direction, δ is the maximum difference between initial fibre stresses if the sequential failure process is to occur, D is a constant relating laminate stress to fibre stress and s the fibre spacing.

For large notches, the stress field is extensive and the stress gradients will be low, thus failure occurs when the stress at the notch tip reaches the unnotched strength of the material. This is equivalent to the characteristic distance models with a zero distance. For this case the notched strength is simply determined from the unnotched strength and calculation of the elastic stress concentration factor at the notch. For smaller notches, initial failures remain contained and a damage zone develops. The requirement then is to predict notched strength allowing for this apparent non-brittle behaviour associated with the damage zone. Potter does this by allowing the geometry of the notch to change such that the stress concentration at the tip increases in relation to the effect of the damage zone, until the modified notch satisfies the propagation criterion. Typical results are shown in the figure 11.10 in which the ratio of notched to unnotched strength, R, is plotted versus the radius of a circular notch. The results are compared with predictions using the average stress criterion for a characteristic distance of 2.4 mm for a standard CFRP. Clearly the predictions are good. The average stress predictions fit well where the characteristic distance is close to the experimental value, that is where the fit was made, but less good at shorter and longer cracks, whereas the Potter model describes behaviour well over all crack lengths.

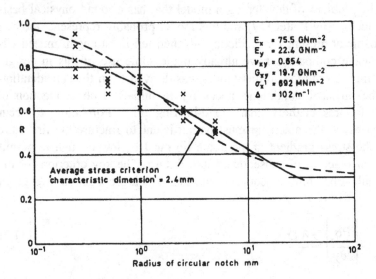

Figure 11.10. Plot of the ratio of notched to unnotched strength versus the radius of a circular notch.

Lastly in this section, mention should be made of damage mechanics type models. The approaches discussed so far have not specifically considered the details of the damage produced at the notches and tried to relate this directly to strength. This generally becomes almost impossible to perform analytically. However, the advent of easy to use finite element analyses has made this approach feasible. Recent work by Beaumont et al. [2] is typical and shows some success in simple laminate configurations under tensile loading. Figure 11.11 shows predictions against experiment, for the range of cross-plied GRP studied, demonstrating good agreement.

The method, however, incorporates various assumptions, based on which damage mechanisms dominate the failure process. These must be known and thus a large amount of mechanics studies are required to establish the damage processes before strength predictions can be made. For more practical laminate configurations the identification of damage processes may be more complex, and indeed different for different lay-ups. This will limit the utilisation of the technique as a design tool. Nevertheless, this technique has promise, particularly

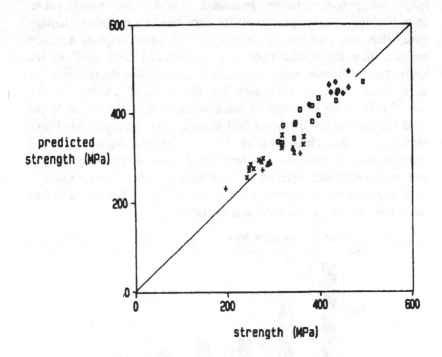

Figure 11.11. Plot of predicted strength versus experimentally derived values for a range of cross-plied GRP.

since it is physically based, something that the author personally applauds, and could yield useful design tools in the future.

11.4.2 Notches - Compression

Turning to compressive loading, which in aerospace is more often design critical, there has been very little work on predictive modelling. Perhaps this is because the challenge of tensile loading was so great that the potentially more complex case of compression seemed beyond achievement. Perhaps the whole problem of performing compression testing has been the deterrent. This review will be confined to the recent work by Fleck et al. at Cambridge University [20], since this seems to combine all the important attributes of a predictive model, a physical base, it is relatively simple, the input parameters are potentially materials properties and the technique gives good agreement with experiment.

Fleck has assumed that compressive failure is dominated by fibre microbuckling from the notch and verified this experimentally for a range of polymer based laminates. Unlike the tensile case, microbuckles usually grow without initiating a significant damage zone, thus the problem is surprisingly in some respects actually simpler than the tensile case and does indeed lend itself to the application of fracture mechanics. Fleck has modelled the microbuckle as a crack, making allowance for the material crushed in the microbuckle zone, and applied fracture mechanics. The inputs to the model at present are the unnotched strength plus a fracture toughness measured on the actual laminate. However, current work is aimed at reducing these requirements to the unidirectional compressive strength and toughnesses measured on unidirectional material. The comparison with experiment is shown in the figure 11.12 for six different laminates and as can be seen is remarkably good.

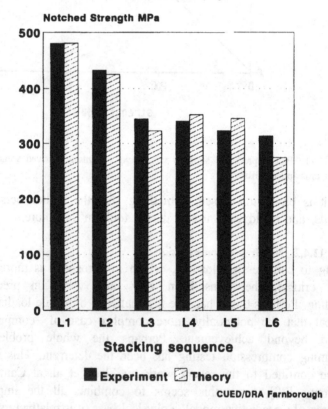

Figure 11.12. Notched compressive strength comparison of theory and experiment CFRP (0m, 45n, 90p) w=25mm r=2.5mm.

11.4.3 Impact

Predictive modelling of impact damage falls into three categories. Firstly how much and what type of damage is produced in the impacting event, secondly the effect this has on mechanical properties and thirdly whether and at what rate this will grow under subsequent loading. The first of these, predicting the damage produced, is very much in its infancy. Researchers have usually been content to simulate damage, either by attempting to duplicate the impact event in the laboratory or by the use of artificial damage in coupons, such as the inclusion of films to simulate delamination. There has been a considerable amount of work, using instrumented impactors, aimed at trying to relate damage events to load/time history and the incident energy, but any links have in general been fairly qualitative. It seems the technique can perhaps rank materials in terms of damage susceptibility, but not yet adequately predict damage type and extent.

Determination of the damage state is usually done experimentally and the term Characteristic Damage State has been used in connection with work of this type [15]. Impact damage is really a combination of three types of damage, namely fibre fractures, delaminations between layers and through-thickness splits along the fibres, which make up the characteristic damage state. In attempting to predict the effect of impact damage on mechanical properties it is necessary to predict the effects of all three types of damage and their interaction. As yet, this has not been satisfactorily achieved, but considerable work is in progress in Europe and the USA. There has been an enormous amount of qualitative work describing the type and extent of damage from experimental observations, but very little predictive modelling.

There has been some good work on trying to predict the effect individually of the three types of damage on residual strength and their rate of growth. In general, small damage areas are often dominated by fibre damage and it is possible to make predictions of residual compressive strength based on limited fibre damage using the notched based models. Larger areas of damage tend to be dominated by matrix cracks and delamination and then sub-laminate instability dominates behaviour. A good example of this is the work in the UK at Cambridge Consultants [21], in which they have written their own FE model, which uses toughness inputs to predict delamination growth. Typical results are shown in figure 11.13 and good agreement was obtained with experimental data from DRA Farnborough. This type of analysis can also be used to predict the stress at instability and thus the residual compressive strength. Typical data from Ilcewicz et

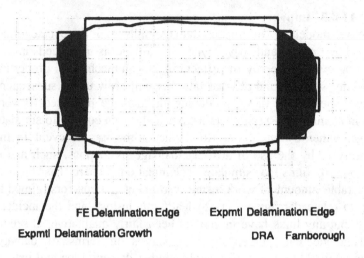

FE Delamination Edge   Expmtl Delamination Edge

Expmtl Delamination Growth

DRA   Farnborough

Figure 11.13. Prediction of delamination growth; FE versus experimental results.

al. [22] is shown in the figure 11.14.

Prediction of damage growth, statically and in fatigue, has generally been by the application of fracture mechanics. Delaminations, as just shown, can be coped with well, but complex areas of damage are more difficult to deal with. Some work has been done trying to develop analogies between impact damage and controlled stress concentrations, such as holes, with some success. However these analogies are usually laminate specific and are unlikely to lead to useful predictive design tools. The work in progress at present, generally based on the development of FE models, seems much more promising.

Finally in this section a mention of continuum damage mechanics, which perhaps should have been discussed earlier in this review. Continuum mechanics has the potential for the prediction of the initiation and growth of damage in composites [23]. The technique is based on the four major laws of conservation of mass, energy, linear and angular momentum with appropriate boundary conditions applied. Energy principles are particularly useful for developing criteria for fracture processes in composites. Realistic models describing undamaged materials have been proposed, but there are considerable difficulties in the development of constitutive relations for the case when the composite is damaged. In general, many parameters have to be introduced that need to be measured experimentally, and models

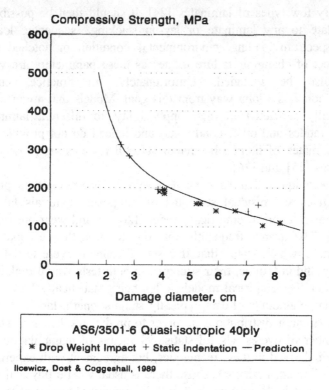

Figure 11.14. Theory versus experiment for CAI (theory based on sublaminate stability and maximum strain criterion).

tend to be specific to particular material combinations. Current work is trying to link the continuum approach to events at the microstructural level and this may lead to greater success.

11.5 FATIGUE AND ENVIRONMENTAL EFFECT

11.5.1 Fatigue Life Prediction
The most widely applied technique for the life prediction of metallic components is linear elastic fracture mechanics. This technique is based upon the growth of single flaws to failure, something not observed in composites (except perhaps for the growth of interlayer delamination), thus the applicability of such a technique to composite materials must be considered at the very least to be very restricted.

The ideal life prediction technique for composites would enable fatigue life and residual strength to be predicted from the minimum amount of data, for example, static strength and perhaps fatigue data on a very few types of laminates [24]. It should then be possible to extrapolate to any laminate or lay-up stacking sequence, loading mode, spectrum loading, environmental condition or notched state. The effect of changing failure modes as these parameters are varied must also be included. Unfortunately, in practice, current understanding is a long way from this goal. Models that are available are usually restricted in their applicability to different laminates, loading modes and other conditions, and indeed do not provide very good estimates of fatigue behaviour. Useful reviews are provided in references [25] and [26].

The techniques that have been developed in attempts to predict fatigue lives and residual strengths of composite materials fall into two major categories, empirical models [26-29] and wear-out models [30, 31]. The former frequently rely on the strength life equal rank assumption, which states that if a set of coupons were tested both statically and in fatigue, their strengths in both tests would rank in the same order. The empirical models, often being statistically based, thus require large amounts of data, typically static strength, fatigue life and residual strength distributions in order to predict residual strengths at any number of cycles. Useful statistical means of manipulating data have been derived in these techniques, but agreement with experiment has been limited, perhaps because there is usually little physical basis for these models. Recent work at Bath University is typical of the best achieved to date [29]. Their work has led to an empirical equation that links static tensile and compressive strength to fatigue life through a power law. The only other input required is a single S-N curve, then a full Goodman or life diagram can be generated for the particular material/laminate combination, see figure 11.15. The major criticism made of this type of model is the lack of a physical basis.

Unlike the empirical models, wear-out models are all physically based, specifically on the reduction of some property, such as stiffness or strength, and the ability to predict its decay during fatigue loading. The major problem with this approach is that the parameter chosen to describe the wear-out is inevitably a function of stress. This necessitates testing at many stress levels to describe its behaviour, reducing the usefulness of the technique in minimising data collection. In addition, such techniques have not been very successful at predicting composite fatigue behaviour. Indeed no particularly successful technique currently exists for the life prediction of composite materials.

Lastly, an alternative empirical model known as Miner's rule

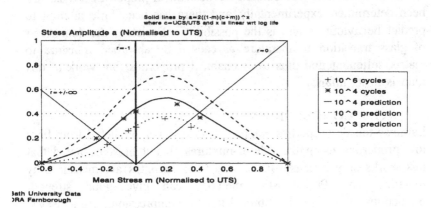

Figure 11.15. Goodman diagram for T800/5245 CFRP (lay-up - 16 ply +/-45.02).

perhaps warrants a mention in its own right. This is used extensively to estimate safe fatigue lives in metals subjected to spectrum loading, with some success. The essence of the model is the Miner sum, i.e. the sum of various fractions of experienced fatigue cycles to those necessary to cause failure at a particular stress level. Its applicability to composites, however, has been found to be rather limited, the main reason being that it provides unconservative estimates of fatigue lives [32, 33]. There may be scope, however, for empirical modifications to be made to this approach to give better predictions of fatigue life under spectrum loading conditions.

11.5.2 Environmental Effects

The principal environmental effect of interest in composite materials is that of moisture absorption. There is clearly a need to be able to predict the amount of moisture absorbed by a composite in a particular environment and also what effect that has on mechanical properties. The absorption of moisture in composites is usually a diffusion controlled process, thus Fickian diffusion theory has been applied with good results. The prediction of moisture content as a function of time and distance through the material is fairly well described by Fickian theory [34]. There can be temperature limits outside of which the behaviour departs from true Fickian and alternative approaches must be sought.

The effect of moisture content on mechanical properties has usually been determined experimentally and there has been little attempt to predict behaviour. There is the possibility of relating the depression of glass transition temperature as caused by absorbed moisture to matrix softening and thus to mechanical properties, but work of this kind is in its infancy.

11.6 CONCLUSIONS

Laminae tensile properties can be reasonably well predicted from fibre tow properties using the rule-of-mixtures, but it must be noted that this works only because of the way fibre strengths are measured by manufacturers. Probabilistic methods can give good strength predictions from single fibre data. In compression, Budiansky's expression gives good agreement with experiment, but needs knowledge of fibre misalignment, a difficult parameter to measure experimentally.

The maximum stress or strain criteria work quite well for simple laminae for simple loading cases, but complex loading requires the use of interaction terms which are often fairly arbitrary. The Tsai/Hill approach is widely used, but this has little physical basis. Hart-Smith advocates the use of a modified Tresca criterion which looks promising.

For plain laminates the first or last ply failure approaches are well used, but they do not allow for ply interactions and delaminations. Mechanistic models look promising and do permit laminate damage to be predicted, but much further work is needed.

The point and average stress criteria are widely used for notched tensile strength, but these require the knowledge of a characteristic distance, experimentally determined, which is not a true materials parameter. Potter has overcome this by basing his model on the stress gradient at the notch tip and obtains good agreement with experiment. Damage mechanics models also look promising for predicting notched tensile strength. In compression, the model proposed by Fleck et al. at Cambridge appears to give good predictions of notched strength but requires further work to reduce the input parameters to materials constants.

Predicting properties in the presence of impact damage is the subject of much on-going research. At present, the prediction of delamination growth seems possible, but the more complex case of delamination combined with fibre fractures and splitting along fibres, as in impact damage, is a formidable problem. Much further work is required.

In general, more effort needs to be devoted to the development of models dealing with design critical notched and damaged cases.

Empirical techniques are beginning to prove useful for the prediction of fatigue life, but there is a need to develop physically based models which do not rely on the generation of large amounts of data as at present.

In general there has been a lot of good work carried out on strength predictive modelling of composites, but there is a long way to go, particularly in developing physically based models and coping with complex stressing situations. In addition, there is a need to put the many individual models together into a framework which can be used for design purposes to avoid designers having to build their own design tools based on a multitude of small models.

11.7 ACKNOWLEDGEMENTS

This work was carried out with the financial support of D. Science (Air), MOD and the DTI Air Division.

11.8 REFERENCES

1] Tsai, S.W., Wu, E.M. "A General Theory of Strength for Anisotropic Materials", J. Comp. Mater., 5, 1971. p 58.

2] Kortschot, M.T., Beaumont, P.W.R., "Damage Mechanics of Composite Materials: 1 - Measurements of Damage and Strength, 2 - A Damage Based Notched Strength Model", Comp. Sci. Tech. 39, 1990. pp 289-326.

3] Guild, G.J., Ogin, S.L., Smith, P.A., "Modelling of 90° Ply Cracking in Crossply Laminates", J. Comp. Mater., (to appear).

4] Zweben, C., Rosen, B.W., "A Statistical Theory of Material Strength with Application to Composite Materials", J. Mech. Phys. Solids, 18, 1970. p 189.

5] Curtis, P.T., "A Computer Model of the Tensile Failure Process in Unidirectional Fibre Composites", Comp. Sci. Tech., 27, 1986. p 63.

6] Curtis, P.T., "The Effect of the Interface on Carbon Fibre Composite Properties, RAE TM Mat 1166, 1991.

7] Curtis, P.T., Gates, J., Molyneux, C.G., "An Improved Engineering Test Method for the Measurement of the Compressive Strength of Unidirectional Carbon Fibre Composites", Composites, **22** (5), 1991. p 363.

8] Rosen B.W., "Mechanics of Composite Strengthening - Fiber Composite Materials", Am. Soc. Metals, 1964. pp 37-45.

9] Jelf, P.M., Fleck, N.A., "Compression Failure Mechanisms in Unidirectional Composites", J. Comp. Mater., (to appear).

10] Budiansky, B., "Micromechanics", Computers & Struct., **16**, (4), 1983. pp 3-12.

11] Tsai, S.W., "Composites Design", AFWAL-TR-84-4183, 1983.

12] Hill, R., "The Mathematical Theory of Plasticity", OUP, London, 1950.

13] Hart-Smith, L.J., "A Strain-based Maximum Shear Stress Failure Criterion for Fibrous Composites", Proc. AIAA/ASME Conf. *Structural Dynamics and Materials*, California, April 1990.

14] Jameson, R.D., "The Role of Microdamage in Tensile Failure of Graphite/ Epoxy Laminates", Comp. Sci. Tech., **24**, 1985. pp 83-99.

15] Reifsneider, K.L., Henneke, E.G., Stinchcombe, W.W., "Defect-Property Relationships in Composite Materials", Proc. 14th Annual Soc. Engg. Sci. Meeting, Lehigh University, PA, 1977.

16] Waddoups, M.E., Eisenmann, J.R., Kaminski, B.E., "Macroscopic Fracture Mechanics of Advanced Composite Materials", J. Comp. Mater., **5**, 1971. pp 446-454.

17] Whitney, J.M., Nuismer, R.J., "Stress Fracture Criteria for Laminated Composites Containing Stress Concentrations", J. Comp. Mater, **8**, 1974. pp 253-265.

18] Awerbuch, J., Madhukar, M.S., "Notched Strength of Composite Laminates: Predictions & Experiments", J. Reinf. Pl. Comp., **4**, 1985. p 3.

19] Potter, R.T., "Tensile Notch Sensitivity in Fibre Composite Laminates", RAE TR 89028, 1989.

20] Fleck, N.A., "Prediction of Notched Compressive Strength of Carbon Fibre Reinforced Plastic", Final report under MoD/SERC Research Agreement GR/F56963, 1992. N.A.Fleck.

21] Pavier, M.J., Johnson, G.E., "Finite Element Modelling of Artificial Damage in Carbon Fibre Epoxy Laminates", CCL final report under MoD Contract SLS41b/1266, March 1991.

22] Ilcewicz, L.B., Dost, E.F., Coggeshall, R.L., "A Model for Compression after Impact Strength Evaluation", Proc. 21st Intl. SAMPE Tech. Conf. September 1989.

23] Talreja, R., "A Continuum Mechanics Characterisation of Damage in Composite Materials", Proc. Roy. Soc. Lond, A399, 1985. p 195-216.

24] Curtis, P.T., "A Review of the Fatigue of Composite Materials", RAE TR 87031, 1987.

25] Gordon, B., Whitehead, R.S., "A Critical Review of Life Prediction Methods for Composite Structures", BAe(MAL) Report SON(P) 179, 1979.

26] Barnard, P.M., Young, J.M., "Cumulative Fatigue and Life Prediction of Fibre Composites", Final Report under MoD Agreement 2028/0134 XR/MAT, 1986.

27] Hahn, H.T., "Fatigue Behaviour and Life Prediction of Composite Laminates", AFML-TR-79-43, 1978.

28] Yang, J.N., Du, S., "An Exploratory Study into the Fatigue of Composites under Spectrum Loading", J. Comp. Mater., 17, 1977. p 511.

29] Adam, T., Gathercole, N., Harris, B., Reiter, H., "A Unified Model for Fatigue Life Prediction Of Carbon Fibre/Resin Composites". Proc. 5th Eur. Conf. Comp. Mater. Bordeaux, April 1992.

30] Pipes, R.B., Kulkarni, S.V., McLaughlin, P.V., "Fatigue Damage in Notched Composite Laminates", Mats. Sci. Engg., 30, 1977. p 113.

31] Poursatip, A., Ashby, M.F., Beaumont, P.W.R., "Damage Accumulation during the Fatigue of Composites", Proc. 4th Intl. Conf. Comp. Mater., Tokyo, 1982.

32] Rosenfeld, M.S., Gause, L.W., "Compression Fatigue Behaviour of Graphite/ Epoxy in the Presence of Stress Raisers, ASTM STP 723, 1981. p 174.

33] Heath-Smith, J.R., "Fatigue of Structural Elements in Carbon Fibre Composite - Present Indications and Future Research, RAE TR 79085, 1979.

34] Shen, C.H., Springer, G.S., "Moisture Absorption and Desorption of Composite Materials, J. Comp. Mater., 10, 1976. p 2.

12 CONSIDERATIONS OF FAILURE THEORIES IN DESIGN

12.1 INTRODUCTION

The physical basis of failure theories has been discussed in Chapter 11, and a more general summary is also provided by Raghava [1]. The purpose of this Chapter is to suggest how failure theories may be used and the extent of their relevance in the design of conventional ship structures.

This Chapter also includes a discussion of the practical modes of failure that may be expected in FRP marine structures and how the designer may avoid them.

12.2 FAILURE THEORIES AND CRITERIA

When the procedures for structural design of FRP vessels were developed in the 1960s and 1970s the science of failure of composites was in its infancy and was not, in general, acknowledged by ship designers. The failure of FRP structures was understood in a very simplistic manner and on the whole viewed uniaxially in compression and tension.

This view was not, and indeed is not, as blinkered as it may seem. Irrespective of the material of which a ship's hull is fabricated the predominant stresses in most of the primary structure are uniaxial. Even when in-plane compression and tension are superimposed on lateral loads, the direction of stiffening ensures that the significant stresses are still in one direction.

In addition to the loading being mainly uniaxial, critical modes of failure frequently involve elasto-plastic buckling due to compressive wave loading rather than material failure, and so again, traditional failure theories are inappropriate even for metal structures. For FRP where the stiffness is much less and no yielding can exist, failure tends mainly to be by elastic buckling and much of the development effort has gone into an understanding of that phenomenon.

It is not surprising therefore, that failure theories have been given relatively little attention by ship designers, whatever the material used.

Having said that however, for metal structures it has become commonplace to apply the Von Mises criterion which is suitable for isotropic materials exhibiting a clear yield point and with a significant post-yield capability. This is clearly not true for FRP, and where any criterion was needed it has become the habit to use the Tresca or maximum stress criterion illustrated in figure 12.1, (although designers may not realise that is what they are doing!).

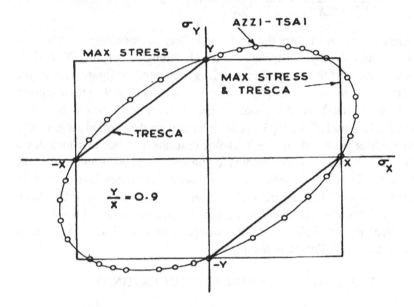

Figure 12.1. Principal failure criteria.

In practice, because of the cyclic nature of loads on a ship and because if any compression existed, failure in buckling was most likely, it was only the tension-tension quadrant that was relevant. Furthermore, as virtually all high quality FRP hulls are built of woven roving laminate with the rovings aligned with the principal stresses, there is little interaction between principal axes and the assumed independence of strength in the x and y directions is reasonable. The criterion is also particularly easy to use, being a direct comparison between the principal stress and material uniaxial strength, with a suitable stress factor.

It may, in the future, be more appropriate to use the Hart-Smith criterion [2] which is based on failure strain, but currently there is insufficient published data on failure strain of marine composites for its use to be feasible.

The above discussion applies to the majority of ship structures. However, there will be areas of structure for example in machinery seatings or in the region of holes where complex stress patterns exist. For such areas Smith [3] recommends use of the Azzi-Tsai criterion [4] thus

$$\frac{\sigma_x^2}{X^2} - \frac{\sigma_x \sigma_y}{X^2} + \frac{\sigma_y^2}{Y^2} + \frac{\tau_{xy}^2}{S^2} = 1. \tag{12.1}$$

where σ_x, σ_y and τ_{xy} are the direct and shear stresses respectively, and X, Y and S are composite uniaxial strengths in the X and Y directions, and the shear strength. It is illustrated in figure 12.1 in the absence of shear stress ($\tau_{xy} = 0$) and for Y/X = 0.9. This criterion reduces to the Von-Mises one in the isotropic case where X = Y = yield stress and $S = X/\sqrt{3}$. A more complete description of laminate strength is provided by the Tsai-Wu criterion [5], but this includes a factor that has to be determined empirically and so is more difficult to use, and is only slightly more accurate than Azzi-Tsai. It will be seen however, from figure 12.1, that in the absence of shear the Azzi-Tsai criterion is close to Tresca and never more than 20% different, so that in the light of safety margins described elsewhere it is suggested that Tresca is adequate.

12.3 FAILURE IN SINGLE SKIN LAMINATES

12.3.1 Failure Modes
The failure of single skin laminates, stiffened and unstiffened, has been discussed at some length. As stated above, the majority of failure modes include an element of buckling instability in marine structures. However, in whatever way the failure is caused, the local effect is cracking of the surface resin followed by failure of the fibres in the outer ply (note that due to the inevitable presence of some flexure an outer ply will always fail first unless there is such severe delamination in the laminate that the strength would be degraded anyway). The outer ply failure is usually known as "First-Ply-Failure" or FPF.

Because of the wet environment, any resin cracking will allow water ingress to the laminate which is likely to cause further, possibly rapid, degradation of strength and stiffness. Consequently such failure should be avoided and a significant margin on FPF is required. Smith [3] recommends that for uniaxial or flexure stressing the design stress should not exceed 30% of the yield stress and preferably should be around 20%. This implies that the figure on the right hand side of equation (12.1) should not be unity but be between 0.04 and 0.09, say 0.05!

The designer must also be aware that because of the low stiffness of FRP there is very likely to be a limit on the allowable deflection of the structure particularly for decks and bulkheads under lateral load. The limit cannot of course be defined absolutely but must depend on the particular application. As an example in the FRP deckhouse designs described in reference [6] deflection limits for stiffeners of 25 mm and plating between stiffeners of 10 mm were assumed.

12.3.2 Environmental Effects

Strength and stiffness are also affected by the environment, in particular humidity and temperature. Most resins reach an equilibrium state for water absorption although the effect on properties may not be fully apparent for some years. Water absorption will occur much more rapidly if any fibre reinforcement is exposed and so it is important that any edges cut during fabrication are coated with resin, and that any damage in service is similarly made good as soon as possible. Once water has entered a laminate it is very difficult to remove it, or indeed to measure how far it has penetrated.

Smith [3] suggests that stiffness and tensile strength of polyester resin can be reduced by up to 30% after many years of immersion in sea water, while flexural and compressive strength may be reduced by up to 35%. However, not all of a ship's structure is immersed and much of the structure will absorb far less water, so it is suggested that an average reduction of 15% of strength and stiffness is used to allow for water absorption. The effect will be less on epoxy and vinyl ester resins, and probably more on phenolic resins, but the latter have not been studied in any great detail in the marine context.

The effect of high temperature on laminates is very marked and so they should not be used in areas where elevated temperatures are likely. Polyester resins are particularly susceptible, losing strength steadily above about 50°C and by 150°C reducing to 10% of strength and 50% of stiffness [3]. Phenolic and vinyl ester resins are less susceptible, but are still down to 50% and 70% of strength respectively and about 80% of stiffness by 150°C.

It is interesting to note, however, that at low temperatures the strength and stiffness of laminates is maintained and in some cases is increased.

12.3.3 Jointing

Jointing FRP unless some form of mechanical attachment is used, will usually introduce a weakness. A "secondary bond" between two pieces of FRP is never as effective as part of the primary lay-up if the same resin is used. If, however, an adhesive that is tougher or more compliant than the matrix resin is used then it is possible that the strength of the joint may be maintained, but probably at the expense of stiffness.

If mechanical joints are specified then through-bolting is the most effective and indeed the integrity of the joint can be maintained even if the bond line fails completely. However, there is a risk of "pull-through" failure of the bolts if large enough washers are not used to spread the load into the laminate. Generally a diameter of washer 2.5-3 times that of the bolt will be needed.

There is no definite rule for failure criteria at joints. In practice, provided that a good joint is made either by using a tough adhesive bond or mechanical fasteners, then no specific allowance needs to be made in the design.

12.3.4 Scaling Effects

In developing standards for design and fabrication of FRP it will not always be possible to use full-scale samples or experimental models. However, it is difficult to scale FRP because to be realistic the same number of plies should be used in the model as in the full-scale leading to a requirement for very lightweight reinforcement. Even if the requisite weight of reinforcement is available, the fibre diameter will not be scaled and results may not be reliable. Moreover, such factors as void content and other fabrication effects cannot be scaled.

For practical purposes therefore, except for very small specimens for measuring simple material properties, it is not realistic to test at less than half scale. It is worth noting that of all the mine countermeasures vessels constructed worldwide, where scale prototype testing has been undertaken, none have been tested at less that half scale and most at larger scale.

12.4 SANDWICH STRUCTURES

12.4.1 Modes of Failure

Sandwich structures suffer similar failure modes to monocoque structures, that is First-Ply-Failure and elastic buckling but in addition there also exist peculiar failure modes relating to buckling of the skins, rupture of the core and separation of the skins from the core.

A discussion on how the failure loads may be predicted is covered in a separate Chapter 11 and references [3] and [7] and therefore, does not merit repeating here. However, it is equally important that water is not allowed to enter the laminate, in fact probably more so because the skins are thinner than a monocoque panel, and the bond line between skin and core may be especially susceptible to water degradation.

Another point to be noted is that if the core, or core to skin bond, fails then even if the skins remain intact the stiffness of the sandwich will very rapidly reduce because the two skins are then acting as independent panels. It is, therefore, essential that the integrity of the core is maintained under all conditions.

It is suggested, therefore, that the Tresca failure criteria is appropriate to sandwich panels in the same way as for monocoque panels, but equally a large margin is required between design load and failure load. The Tresca criterion does not, however, take account of shear effects which are important in the core and for this it is suggested that the Azzi-Tsai criterion is used, but with the factor on the right hand side of equation (12.1) being 0.3 or less if there is uncertainty over the core shear strength.

Because, also, of the criticality of skin buckling and the potentially catastrophic effect it has on the overall stiffness of the structure, it is suggested that a safety factor of at least 4 is used between design load and buckling load.

12.4.2 Joint in Sandwich Panels

Joining sandwich panels is not straightforward because of the difficulty of ensuring continuity of the core. A suggested procedure which should lead to the strength being nearly the same as a parent panel is illustrated in figure 12.2. The core is left proud on both panels and bonded at the edge and then bridging sections of skin are laid up in situ. The precise dimensions of the joint will probably need to be developed by trials for a specific configuration.

Tee butts and other joints are probably best made in a similar way to monocoque structures with no attempt being made to join the cores. A tee butt is illustrated in figure 12.3 in general terms but again detailed dimensions would probably result from trials. However, as there is, in effect, a gap in the joint the flexural strength must be estimated as the sum of the tensile or peel strength of one boundary angle and the compression strength of the other. Because of the

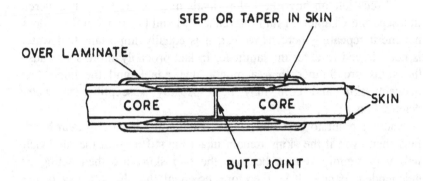

Figure 12.2. Typical joint in sandwich panel.

uncertainties involved a large safety margin is recommended.

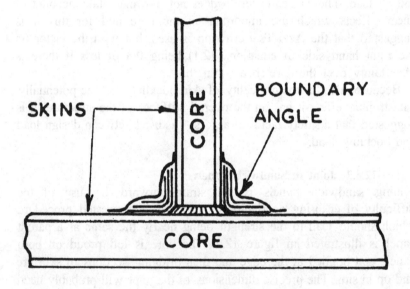

Figure 12.3. Typical tee butt joint in sandwich.

12.5 REFERENCES

1] Raghava, R.S., "Prediction of Failure Strength of Anisotropic Materials", Lee, S.M., (ed.), *Reference Book for Composites Technology*, **2**, Technomic, Lancaster, 1989.

2] Hart-Smith, L.J., "A Strain-based Maximum Shear Stress Failure Criterion for Fibrous Composites", Proc. AIAA/ASME Conf. *Structural Dynamics and Materials*, California, April 1990.

3] Smith, C.S., "Design of Marine Structures in Composite Materials", Elsevier, London, 1990.

4] Azzi, V.D., Tsai, S.W., "Anisotropic Strength of Composites", Exp. Mech., **5**, September 1965. p 283.

5] Tsai, S.W., Wu, E.M., "A General Theory of Strength for Anisotropic Materials", J. Comp. Mater., **5**, January 1971. pp 58-80.

6] Smith, C.S., Chalmers, S.W., "Design of Ship Superstructures in Fibre-Reinforced Plastic, Trans. RINA, **129**, 1987. pp 45-61.

7] Olsson, K-A., Reichard, R.P., (eds.), "Sandwich Construction I", EMAS UK, 1989.

13 TEST PROCEDURES AND STANDARDS

13.1 INTRODUCTION

Standardised test methods are needed to characterise the properties of laminates and sandwich materials at several stages in the product design and production cycle. These stages include the initial design and material selection, the detailed design, the development of the production process and the evaluation of the final product.

Test methods may characterise material fabricated directly as test panels from fibres and matrices or from pre-impregnated material (prepregs), as well as that taken from a manufactured product. Test panels should represent closely the final production conditions. For example, vertically and horizontally produced material are likely to have different properties due to the differences in resin drainage etc. Similarly, specimens should in all cases adequately represent the structure being tested.

A major difference between the marine use of reinforced plastics and the use of advanced prepreg materials is involvement in the former case with thicker sections. The need for appropriate test methods for thick sections is recognised and a group exists in the USA for this purpose related to the use of filament wound cylinders for submersibles. For both application areas there is increased interest in the through-thickness properties, which may be the performance limiting parameter. In addition the marine industry is more likely to be using mat and woven material formats although fabrics are used extensively in the aerospace industry.

Recently, a major initiative on European standards (ENs) in support of the European Single Market was launched. EN standards have to be published in the member countries (CEC + EFTA) and existing standards of the same scope withdrawn. In addition, once work has started in CEN a stand-still arrangement exists that precludes work on a national standard of the same scope; although work is possible, and necessary, in support of the draft development and national vote on the European standard.

General aspects of the development and validation of test methods are covered in section 13.2, with the remaining sections covering test methods for individual properties. In each case available international standards are listed. The code used for these standards are; * = draft or work/study item, R = standard under revision, A = Aerospace series and GEN. = CEN general engineering series. It should be noted that the "alternatives" listed are not necessarily technically equivalent.

13.2 STANDARDS DEVELOPMENT AND VALIDATION

13.2.1 Standards Organisations

The standards with the widest application are the world-wide ISO (International Standards Organisation) series and the CEN (Comité Européen de Normalisation) series for use in CEC and EFTA countries [1]. The ISO series are fairly comprehensive for glassfibre reinforced plastics (GRP) but only in 1990 did significant work begin on test methods for carbon fibres, carbon fibre reinforced plastics (CFRP) and other advanced composites.

The ISO standard making process can be slow as only one meeting to ratify decisions is held each year. The EN series produced by CEN is likely to be quicker as meetings are held more frequently in order to comply with the directive to produce the standards and specifications required in support of the European Single Market. Test method standards are required earliest in order to allow the material specification standards being prepared to call-up appropriate test methods to prove compliance with the specification. The main emphasis in the EN "General Engineering" series is to adopt, where possible, existing ISO test methods. The CEN standards also includes an "Aerospace" series, although some harmonisation may be possible as the general series is developed. In contrast both ISO and ASTM have a single series covering all types of users.

Future work will be shared between ISO and CEN via the Vienna Agreement which will, in areas of mutual interest, allow one of the organisations to develop the standard on behalf of both bodies for parallel vote. There are responsibilities on both sides, for example an ISO draft must be produced within the CEN timescale; while CEN would have to produce a draft acceptable to the wider international community within ISO. Each organisation can produce its own standards in areas where there are no common interests.

In both international series (ISO, CEN) the delegates and voting rights reside in the national bodies; the most active of these with their own national series are AFNOR, ANSI (ASTM), BSI, DIN and JIS.

There are many informal group and company test methods (e.g. as in [2]) in use. However, as many of these methods are related to, or devised from other test methods, there is now a major attempt to obtain international harmonisation, and validation, of standard test methods [3].

The major effort to date has been towards ensuring standard test methods are available for the high performance composites such as for carbon fibre reinforced plastics (CFRP) as existing ISO standards mainly cover glass fibre reinforced plastics (GRP). Some of these methods can be extended to cover other fibre based systems. Harmonised and quicker availability of standards for new test methods should arise in future through the international pre-standards research being under taken through the VAMAS initiative (e.g. fatigue test methods [4,5]). The Versailles Project on Advanced Materials and Standards (VAMAS) programme concerns international cooperation on pre-normative research aimed at the development of recommended procedures. It is based on technical work in the Group of 7 countries (Canada, France, Germany, Italy, Japan, USA, UK and CEC) and in other countries applying to participate in individual programmes.

13.2.2 Validation of Standards

The increased importance of standards as part of the legal framework for free trade agreements and/or product liability has resulted in increased adherence to the requirements for the precision of each test method to be validated by round-robin tests [6]. The precision of the method so established is stated in the standard and has to be commensurate with its intended use (i.e. "fitness for purpose"). Providing the assessment is based on several materials a minimum of eight test laboratories can be used [7]. Increasingly, it is necessary that the test equipment should be calibrated, as by BCS (British Calibration Service) and that in the UK, ideally, the laboratory should have NAMAS accreditation (National Measurement Accreditation Service) or work to ISO 9000/EN 29000/BS 5750 procedures.

Precision is defined in ISO 5725 as "the closeness of agreement between mutually independent test results obtained under stipulated conditions". Two measures of precision, "repeatability - r" and "reproducibility - R", are necessary to describe the variability of a test method.

The repeatability value is defined as the value below which the absolute difference between two single test results obtained under repeatability conditions may be expected to lie within a probability of

95%. Similarly the reproducibility value R also refers to a 95% probability level.

REPEATABILITY CONDITIONS	REPRODUCIBILITY CONDITIONS
same method	same method
identical material	identical material
same laboratory	different laboratories
same operator	different operators
same equipment	different equipment
short interval of time	

One difficulty for reinforced plastics/composites in these validation exercises is related to the freedom of the designer to design the material as well as the structure. Consequently there are few standard products available as this is in conflict with the basic philosophy of composites. In designing a validation exercise a representative range of materials need to be used which normally means the main production materials. The precision clause in the standard will state the materials used in the evaluation. Validation exercise are to be run by NPL on behalf of ISO for the new tensile test (ISO 527-5, see section 13.5) and on behalf of BSI on ten test methods based on the CRAG recommendations [2].

13.3 TEST PANEL AND SPECIMEN MANUFACTURE

The various standards defining this aspect are listed in table 13.1. A major revision is being undertaken of ISO 1268 covering test panel manufacture. The original document for low-pressure moulding of laminated GRP plates is to have its scope widened so that all fibre reinforced materials manufactured by any method are included. The nine parts will include in addition fabrication by injection moulding, filament winding and autoclaves. Where these methods are covered by existing standards they will ultimately be withdrawn. NPL is undertaking a round-robin study on test panel manufacture. Material from a single batch purchased by NPL has been sent to 12 different sites to prepare test panels to be returned to NPL where all the mechanical testing will be undertaken. The opportunity is also being used to compare C-scan procedures for assessing the quality of these test panels.

Although some aspects of specimen preparation are covered by ISO

2818 on machining of plastics and by instructions in some test methods, it is clear that a need is developing for a specialised document to cover reinforced plastics/composites. Aspects to be covered include machining according to the primary fibre directions, and the use of laser/water jet cutting techniques as well as the more usual diamond slitting wheels. Diamond tools while excellent for carbon and glass fibre based material are not suitable for aramid fibre based systems.

Table 13.1. Standard procedures for test panel and specimen manufacture.

NUMBER	TITLE	ALTERNATIVES
ISO 1268 (R)	Plastics-Preparation of glass fibre reinforced, resin bonded, low-pressure laminated plates or test panels	BS 2782/1/920A
EN 2374 (A)*	Glass fibre reinforced mouldings and sandwich composites-Production of test panels	
EN 2565 (A)	Preparation of carbon fibre reinforced resin panels for test purposes	
ISO 2818	Plastics-Preparation of test specimens by machining	
ISO 3597-1	Textile glass-Rovings-Determination of mechanical properties on rods-Part 1: Generalities and preparation of rods	

13.4 COMPONENT FRACTIONS AND THICKNESS MEASUREMENT

Standard test methods for determining constituent contents are listed in table 13.2. The components to be measured include the fibre, void and resin fractions. For production purposes, these fractions are normally calculated as weight fractions. In contrast the theoretical prediction of properties is normally related to the volume fractions of components. Similarly, production will often control the resin content (cf. fibre content) as this is the component that needs most control; the fibre content being controlled through the number of layers specified to be used. The relationships between volume (V_f, V_m) and weight (W_f, W_m) fractions are given by:-

$$V_f = \frac{\rho_c}{\rho_f} W_f \ , \ V_m = \frac{\rho_c}{\rho_m} W_m. \tag{13.1}$$

For the usual production materials based on glass fibre there are well established test methods and corresponding standards using the resin burn-off method. In these methods an oven is used at a temperature of 625°C for a sufficient period to burn-off the resin in an air atmosphere. It is important that the oven fumes are not allowed to enter any inhabited area. A modified method is used if it is necessary to separate the fibre from a filler.

Table 13.2. Standard test methods for the determination of constituent contents.

NUMBER	TITLE	ALTERNATIVES
ISO 1172	Textile glass reinforced plastics: determination of loss on ignition.	BS 2782/10/1002, NF T57-102 ASTM 2584
ISO 7822	Textile glass reinforced plastics- Determination of void content-Loss on ignition, mechanical disintegration and statistical counting methods	ASTM 2374
EN 2564 (A)*	Carbon fibre laminates-Test method for the determination of the fibre and resin fractions and porosity content	ASTM D3171

Loss on ignition (also referred to as resin burn-off) can be used to determine the void content but ISO 7822 suggests the method is only accurate to ±2.5% by volume. The mechanical disintegration method measures the density before and after crushing and is suggested to have an accuracy of ±1%. It is necessary that the material can be crushed either under ambient conditions, or after cooling, to expose all the enclosed voids. It is necessary to know the density of both the fibres and resin. The statistical counting of voids by analysis of a polished cross-section is particularly suitable for low void contents of <1%. This would be the level expected in good quality prepreg and filament wound structures. ISO 7822 contains in a normative annex the procedure reproduced in table 13.3 for polishing sections (see also [8] for an alternative procedure)

Table 13.3. Specimen polishing schedule (Annex A - ISO 7822).		
SPEED	200 rpm	
FORCE	~5 N (500 gf) (for a 25 mm diameter section)	
Polishing Schedule		
TIME(min)	GRAIN	REMARK
3	400 (35μm)	Copious watering
5	800 (22μm)	Copious watering
15	12-H alumina (Al_2O_3)	The alumina is used slightly diluted, with an almost pasty consistency

When, for high performance applications, carbon fibres are used the chemical digestion technique is required to determine the fibre content. These techniques must be used with care because of the acids and/or hydrogen peroxide used. A recent introduction, most suitable for concerns undertaking large numbers of determinations, is a microwave heated digester with automatic pumping of the reagents in a closed system. A memory feature allows routine conditions to be stored for future use.

Thickness is frequently measured using portable thickness monitors although control of thickness starts with the quality control aspects of production. Thickness is often taken as a nominal value based on the number of plies and the ply thickness. A procedure is being developed for unidirectional materials for normalisation of the test results to 60% volume fraction for fibre dominated properties.

13.5 TENSILE PROPERTIES
Standard test methods for determining tensile properties are listed in table 13.4. The ISO standards for measuring tensile properties are being revised. The standard specifically for GRP, ISO 3268, is to be included in a major revision of ISO 527 to cover all unreinforced, filled and reinforced plastics. The new standard consists of five parts, with Part 1 covering general principles and specific parts covering unreinforced plastics and moulding materials (Part 2), films (Part 3) and reinforced plastics (Parts 4 and 5). In 1992, Committee Drafts (CDs) for international ballot will be circulated covering isotropic and orthotropic materials (Part 4) and unidirectional materials (Part 5). Part 4 will effectively reproduce the available specimens and test conditions in ISO 3268, while Part 5 will be suitable for the higher performance, aligned material including unidirectional tape. The revised standards will apply to all "textile" type fibres of small diameter (5-20 microns). The cross-section sizes adopted have been obtained by constructive interaction of representatives of the main standards bodies involved in this area following an initiative in the UK [9]. Bodies involved included ASTM, JIS, CEN, European national standard bodies and ISO. It is hoped that similar harmonisation can be obtained for several more of the most used test methods. A round-robin validation exercise will be undertaken in 1992/3 for Part 5 using a range of materials.

The general aspects for tensile testing are:-

-   The loading train and specimen axes should be well aligned.
-   Specimens are relatively long and slender.
-   Strain is measured over a long gauge length (e.g. 50 mm)
-   End tabs are used to avoid failure in the grips.
-   Grip pressure should only be sufficient to stop the specimen slipping, excessive grip pressure can cause crushing of the specimen especially for fully aligned material.

The test specimens available in Part 4 are shown in figure 13.1. The dumbbell specimen is not suitable for materials with coarse structure. The Part 5 specimens are straight tabbed-specimens with cross-sections of 15 mm x 1 mm and 25 mm x 2 mm for longitudinal and transverse samples, respectively. The specimen sizes given in Part 5 represent the first successful harmonisation of ISO, EN, ASTM and JIS viewpoints; and included a revision of a current EN-Aerospace draft to reduce the specimen width from 16 mm to 15 mm to harmonise with the new ISO standard.

However there are still many aspects that need to be decided. For

Table 13.4. Standard test methods for determining tensile properties.

NUMBER	TITLE	ALTERNATIVES
ISO 527(R) (to be balloted for CEN GEN)	Plastics-Determination of tensile properties. Part 1: General principles Part 2: Test conditions for moulding and extrusion plastics Part 4: Test conditions for isotropic and orthotropic fibre reinforced plastics composites Part 5: Test conditions for unidirectional fibre reinforced plastics composites	ISO 3268 BS 2782/3/320 BS 2782/10/1003 JIS K7054
EN 2561 (A)*	Carbon-thermosetting resin unidirectional laminates-Tensile test parallel to the fibre direction	JIS K 7073
EN 2597 (A)*	Carbon-thermosetting resin unidirectional laminates-Tensile test perpendicular to the fibre direction	
EN 2747 (A)*	Glass fibre reinforced plastics-Determination of tensile properties	JIS K7054
ASTM D 3039	Standard test method for fibre-resin composites	
ASTM D 2585	Preparation and tension testing of filament wound pressure vessels	
ASTM D2990	Apparent tensile strength of ring or tubular plastics and reinforced plastics by split disk method.	

example, the material and shape of the end tab [10], the strain or load levels used in defining the modulus (tangent or chord) and the specimen alignment required. This latter point has been adopted from the current revision of the ASTM method but care is needed in the application of the procedure and in the selection of the allowed strain levels.

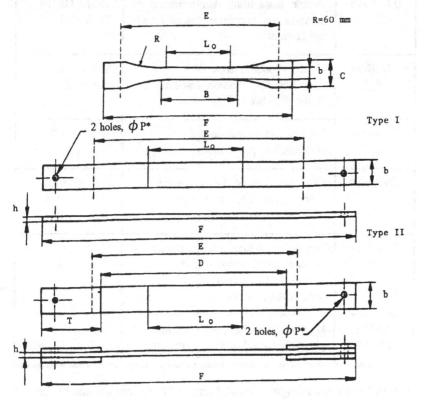

Figure 13.1. Tensile specimens in ISO 527-4 (R).

13.6 COMPRESSIVE PROPERTIES
A list of standard test methods for determining compression properties is given in table 13.5. The main method of testing is to use a strip specimen stabilised by a support jig against Euler buckling. The support jig may be along the full length or more normally provide full support along part of the specimen leaving unsupported a short gauge length. A range of designs are available to support the specimen, see figure 13.2. These include the Celanese jig described in ASTM D 3414 and ISO 8515, Method A, which requires tight control of specimen thickness and is only suitable for 2 mm thick aligned prepreg

Table 13.5. Standard test methods for determining compression properties,

NUMBER	TITLE	ALTERNATIVES
ISO 8515	Textile glass fibre; determination of compression properties parallel to the laminate.	ASTM D3410 JIS K7056
EN 2850 (A)*	Unidirectional carbon fibre laminates compression test-Parallel to the direction of the fibres.	ASTM D695 (modified)
ISO 604	Plastics-Determination of compressive properties.	
EN Gen.*	Determination of the compressive properties of fibre reinforced plastics/composites.	DIN 65380(A) DIN 65375(A)
IS0 3597-3	Textile glass-Rovings-Determination of mechanical properties on rods-Part 2-Determination of compressive strength.	
ASTM D2586	Hydrostatic compressive strength of glass reinforced plastic cylinders.	

material. Several recognised and home-built developments of this jig exist such as the ITTRI test fixture which uses flat wedges rather than conical wedges.

The end loaded specimen support jig given in ISO 8515, Method B is simpler to use and is suitable for material from 2-10 mm thick. Strain gauges on both faces should be used to check that specimen buckling does not occur prior to failure. Development of the test method is underway using both improved jigs [11] and tabbed specimen design; and machined/integral tab specimens [12]. The former developments are more suitable for evaluating standard test panels. Testing of prism specimens, if required, are covered by ISO 604 in the Plastics series but these are not generally suitable for composites.

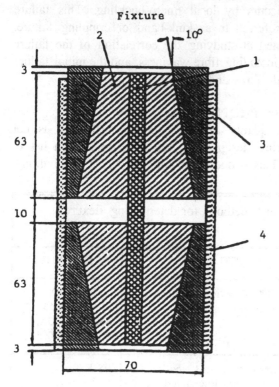

Fixture

1 Specimen
2 Collet grip
3 Tapered sleeve
4 Cylindrical
    shell

(a) Compression fixture for Method A (Celanese)

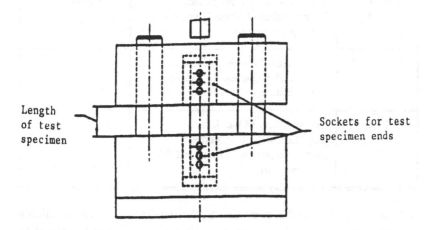

Length
of test
specimen

Sockets for test
specimen ends

(b) Compression fixture for Method B

Figure 13.2.   Compression test method jigs in ISO 8515.

Failure frequently occurs by local micro-buckling. This failure mode is alternatively referred to as kink-band or crippling failure. Current research is aimed at studying the correlation of the failure stress to local structure including fibre waviness, and the initial failure mode of both individual fibres and the composite.

13.7 FLEXURAL PROPERTIES
Although favoured for quality assurance purposes the test is not satisfactory for measuring design data; see table 13.6 for a list of standard test methods. This is due to the complex stress distribution,

Table 13.6. Standard test methods for determining flexure properties.

NUMBER	TITLE	ALTERNATIVES
ISO 178	Plastics-Determination of flexural properties of rigid plastics	ASTM D790, EN63, DIN 53452, BS2782/3/335A, BS2782/10/1005
EN 2561 (A)*	Unidirectional laminates Carbon-thermosetting resin-Flexural test	JIS K7074
EN 2747 (A)*	Glass fibre reinforced plastics-Determination of flexural properties-Three point method.	JIS K7054
ISO 3597-2	Textile glass-Rovings-Determination of mechanical properties on rods-Part 2: Determination of flexural strength.	DIN 53390
EN Gen.*	Determination of the flexural properties of fibre reinforced plastics/composites. (Three and four point methods).	

the mixed failure modes present and the occurrence of early failures on the compressive face associated with stress concentrations due to loading point. Fabric based or thin skinned sandwich laminates are more sensitive to this local loading than mat based materials. It should be noted that if the laminate lay-up is not symmetrical or there is a gel coat on one face; then the strength properties may vary depending on which surface is in tension or compression.

This local loading is reduced in the 4-point loading configuration but the international standard ISO 178 only includes the 3-point version of the test, see figure 13.3. The revised version of this standard is about to be published but will not include 4-point, although this test method will be included in the new CEN general series standard being prepared for composites. The CEN standard will then be made available to ISO as a companion method to the existing standard. It is to be hoped that at this point there can be harmonisation with the two EN-aerospace standards, which use a 2 mm and 3 mm thick specimen for carbon and glass- fibre based laminates respectively.

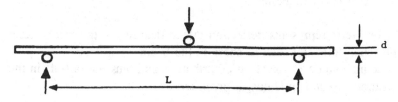

(a). Three-point flexure test for carbon fibre composite (L/d=40).

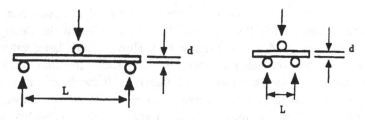

(b). Three-point flexure test for glass fibre composite (L/d=16).

(c). Interlaminar shear strength (L/d=5).

Figure 13.3. Bending and interlaminar shear test configurations.

A major limitation of the test method is that the measured properties are dependent on the stacking sequence tested. Consequently the data obtained can only be applied to other cases if the material is homogeneous through the section (i.e. all one type of reinforcement) or is transversely isotropic (i.e. fully aligned material).

The flexure test is undertaken at a span that will reduce to an acceptable level the through-thickness shear stresses and deflections. For aligned glass fibres the test span to thickness ratio is 16. Larger ratios are required for carbon fibre composites because the shear modulus is relatively lower for these materials due to the higher Young's modulus. The likely error in flexural modulus can be estimated from equation (13.2) for 3-point loading.

$$w = \frac{PL^3}{48\,D} + \frac{PL}{4Q}. \tag{13.2}$$

where

l = test span, b = beam width, d = beam depth
P = load for displacement w
D = flexural stiffness
E = flexural modulus
Q = shear stiffness
G = shear modulus

The first term represents deflection due to flexure and the second term represents deflection due to shear in the beam.

The magnitude of the shear contribution and thus the fall-off in the measured flexural modulus is given by;

$$\frac{\text{shear deflection}}{\text{total deflection}}\ \% = \frac{G}{E}\ (\frac{d^2}{L^2}) \times 100. \tag{13.3}$$

Analysis of tests at several test spans can be used to obtain a more accurate value of the flexural modulus and a value of the through-thickness shear modulus, see figure 13.4. However this latter value is most accurate for materials with a low shear modulus and is particularly appropriate for sandwich materials. It should also be noted that the test span to beam width ratio should be greater than 4, otherwise the test specimen will be closer to a plate geometry and require appropriate plate equations [13].

13.8 INTERLAMINAR SHEAR PROPERTIES
The interlaminar shear strength (ILSS) - see table 13.7 - test is used to measure the strength while the modulus can be obtained from the flexural test described above (also see section 13.9). The ILSS test method is normally not used for design data but for quality control. The test is frequently used as an indirect method of assessing the quality of the interfacial bond. The basis is a 3-point flexure test with a short support span. The shortening of the test span compared to the standard flexure test increases the relative magnitude of the shear stress compared to the flexural stress as noted above. The increased

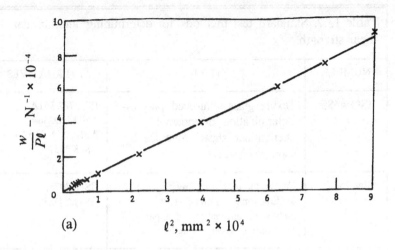

(a)

ℓ^2, mm$^2 \times 10^4$

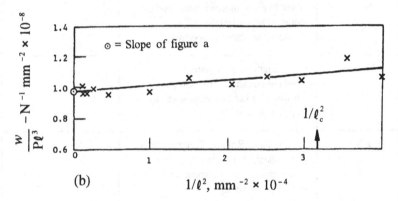

$\odot$ = Slope of figure a

$1/\ell_c^2$

(b)

$1/\ell^2$, mm$^{-2} \times 10^{-4}$

(a)   slope of graph $a = \dfrac{1}{48D} = \dfrac{1}{4bd^3E}$

(b)   slope of graph $b = \dfrac{1}{4Q} = \dfrac{1}{4bdG}$

Figure 13.4. Analysis of flexure tests at different spans.

stress compared to the flexural stress as noted above. The increased shear stress should trigger fracture on, or close to, the neutral plane through an interlaminar shear failure. This requires in most cases a test span to specimen depth ratio of 5. In some cases if the shear strength is high a smaller ratio may be required.

Difficulties in the interpretation of the test arise when a flexural

Table 13.7. Standard test methods for determining interlaminar shear strength.

NUMBER	TITLE	ALTERNATIVES
ISO 4585	Textile glass reinforced plastics-Determination of apparent interlaminar shear properties by short-beam test	BS 2782/341A ASTM D2344 AFNOR NFT57-104 JIS K7057
EN 2563*	Carbon-thermosetting resin unidirectional laminates-Test method-Determination of apparent interlaminar shear strength	JIS K7078, ASTM D2344
EN 2377	Glass fibre reinforced plastics-Test method-Determination of interlaminar shear properties	
EN Gen.*	Determination of apparent interlaminar shear strength of reinforced plastics/composites by short-beam test	
ISO 3597-4	Textile glass-Rovings-Determination of mechanical properties on rods-Part 4: Determination of apparent interlaminar strength.	

failure occurs on the tensile or compressive face of the beam, or there is a mixed flexural/shear failure. In some cases careful fractographic examination is needed to identify the initial failure as being flexural, particularly from the compression face, which triggered a subsequent shear failure. Only results calculated for shear or mixed failure modes should be reported as ILSS results. It should be noted that the test only measures the shear strength and assumes a parabolic stress distribution based on a isotropic material response. This is one of the test methods where a physio-chemical test method would be preferred for quality assurance (QA) purposes.

The test result is calculated from:-

$$\tau = \frac{3\ F}{4\ b\ d} \tag{13.4}$$

where
- F = failure load, N
- b = specimen thickness, mm
- d = specimen width, mm
- τ = apparent interlaminar shear strength

13.9 IN-PLANE SHEAR PROPERTIES

Several test modes are available for the in-plane shear properties, as can be seen from table 13.8, although few are available as international standards. The most frequently encountered include torsion of a tube or rod, tensile test of ±45° layered material, Iosipescu, rail shear and plate twist. The 2- and 3- rail shear test although acceptable for modulus measurement, are not reliable for strength measurement and require a large, expensive specimen. The plate twist test is conducted in a flexure test mode, uses a very simple specimen but only gives the in-plane shear modulus. Work at NPL has improved the analysis of the plate-twist test and a round-robin is in progress to obtain precision data [14].

Table 13.8. Standard test methods for determining in-plane shear properties.

NUMBER	TITLE	ALTERNATIVES
ASTM D3518	In-plane shear stress-strain response of unidirectional reinforced plastics (±45°).	
ASTM D4255	Standard Guide for testing in-plane shear properties of composites laminates (rail).	
ASTM D3846	Standard Test Method for in-plane shear strength of reinforced plastics (notched beam D690 jig).	
ASTM*	In-plane shear properties of hoop wound polymer matrix composite cylinders.	
ASTM*	Shear properties of composite materials by the V-notched beam method.	

Most use has been made of the tensile test but it is only suitable for materials prepared with ±45° fibre orientation. Recent interest, including an ASTM round-robin, has been shown in the Iosipescu test on a notched beam [15]. Other tests available are torsion of a beam and tension or compression of a notched beam (scissor shear). Not all of these methods are suitable for both shear stiffness and strength measurements.

Depending on the format of the sample available the tube torsion or plate twist are the preferred methods for measuring modulus standard, see figure 13.5. The Iosipescu test is gaining popularity for the measurement of strength properties, in particular.

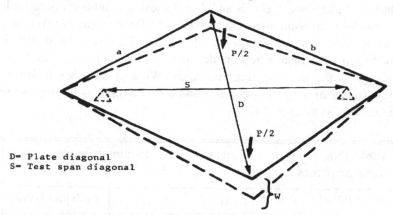

D= Plate diagonal
S= Test span diagonal

(a). Plate twist.

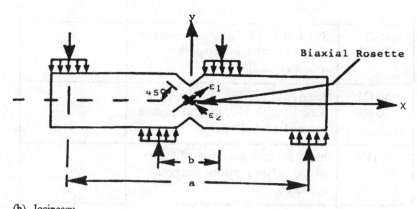

(b). Iosipescu.

Figure 13.5. Plate twist and Iosipescu test geometries.

The modulus is given in the plate twist test by;

$$G_{12} = \frac{3Kab}{8h^3} \cdot \frac{\Delta P}{\Delta w} \qquad (13.5)$$

where

ΔP = applied load increment, N
a   = side length, mm
b   = opposite side length, mm
Δw = deflection for load increment ΔP, mm
h   = plate thickness, mm
K   = calibration factor for loading point position

13.10 SANDWICH MATERIAL PROPERTIES

The ASTM series of tests are the most comprehensive available for sandwich structures, see table 13.9. There is common ground between these tests and laminate tests in the use of a "sandwich beam" geometry for measuring the compression strength of a laminate when used as the weaker skin of a specially fabricated sandwich beam. The configuration of the transverse tension test of sandwich materials is also used for measuring the through-thickness tensile strength of laminates but it is not an easy or very reliable test method.

13.11 ENVIRONMENTAL AND LONG-TERM TEST CONDITIONS

13.11.1 Environmental

There are several test methods available which are generally applicable to plastics, both reinforced and unreinforced as can be seen from table 13.10. One of the most significant factors is that conditioning in the absence of stress normally gives an upper bound result. This area is still developing but already many tests exist with some overlap. The CEN working group on engineering use of composites has initiated a study item to identify the minimum coverage necessary to account for the effect of both vapours (e.g. moisture) and liquid (e.g. chemicals, water) environments. In addition, a standard is required for assessing the amount of take-up of the environment at equilibrium. There is also a need for standard "short-time" treatments that are more applicable to industrial timescales and costs. The difficulty in such tests is to select a time that gives a correct indication for all material types.

Table 13.9. Standard test methods for determining the properties of sandwich materials.

NUMBER	TITLE	ALTERNATIVES
C 271	Density of core materials for structural sandwich construction.	
C 272	Water absorption of core materials for structural sandwich construction	
C 273	Shear properties in flatwise plane of flat sandwich constructions or sandwich cores	
C 297	Tensile strength of flat sandwich constructions in flatwise plane	
C 364	Edgewise compressive strength of flat sandwich constructions	
C 365	Flatwise compressive strength of sandwich cores	
C 366	Measurement of thickness of sandwich cores	
C 393	Flexural properties of flat sandwich constructions	
C 394	Shear fatigue of sandwich core materials	
C 408	Flexural-creep sandwich constructions	
C 481	Laboratory ageing of sandwich constructions	

13.11.2 Fatigue and Creep

There are few standards available for reinforced plastics or composites for long-term loading, see table 13.11. In general any test configuration used for determining static properties can, and normally has been used under fatigue loading conditions. Consequently, as the static test methods are only now being agreed upon, it would have

Table 13.10. Standard test methods for determining
environmental resistance.

NUMBER	TITLE	ALTERNATIVES
ISO 291	Plastics-Standard atmospheres for conditioning and testing	BS 2782/0
ISO 62	Plastics-Determination of water absorption	BS 2782/4/430A-D DIN 16946/1, DIN 53495
ISO 175	Plastics-Determination of the effects of liquid chemicals including water	BS 2782/8/830A BS 4618, DIN 53476
ISO 483	Plastics-Small enclosures for conditioning and testing using aqueous solutions to maintain relative humidity at constant value	BS 3718
EN 2378 *(A)	Fibre reinforced plastics-Determination of water absorption by immersion in demineralised water	
EN 2489 *(A)	Fibre reinforced plastics-Determination of the action of liquid chemicals	
EN 2823 *(A)	Fibre reinforced plastics-Determination of the effect of exposure to humid atmosphere on physical and mechanical properties	
ISO 4607	Plastics-Method of exposure to natural weathering	BS 2782/5/550A
ISO 4611	Plastics-Determination of the effects of exposure to damp heat, water spray and salt mist. (on optical and colour properties)	BS2782/5/551A
ISO 4892	Plastics-Method of exposure to laboratory light sources	BS 2782/5/540B, DIN 53387
ISO 6252	Plastics-Determination of environmental stress cracking (ESC)-Constant tensile stress method	DIN 53449
BS 5480 Part 2	Specifications for glass reinforced plastics (GRP) pipes and fittings for use for water supply and sewerage. Part 2 Design and performance requirements	

Table 13.11. Standard test methods for long term performance.

NUMBER	TITLE	ALTERNATIVES
ASTM D3479	Tension-tension fatigue of orientated fibre, resin matrix composites	
AFNOR T51-120	Bending-fatigue test for reinforced plastics with non-stretch specimens	
ISO 899	Plastics-Determination of tensile creep	DIN 53444
ASTM D2990	Tensile, compressive and flexural creep and creep-rupture of plastics	

been premature to define fatigue tests specimens. The available standards do not contain any specific instructions on test frequency (nb. check for autogenous heating), R ratio (min. stress/max. stress) and stress levels. Research under the VAMAS initiative is aimed at developing agreed procedures suitable as the basis of future harmonised standards. This programme is studying the repeatability of both a tensile test method and a flexure test method; and is conducting an inter-comparison of the two methods. Six materials are being used to represent the range of possible materials. Whereas the tensile test is undertaken under load-control when specimen failure (separation) is normally taken at the failure point, the flexure is undertaken using simple electric motors under displacement control. In this later case most composites progressive soften as a result of the distributed microdamage resulting in a fall off in the load applied to the specimen. As a consequence specimen failure is not reached and alternative failure criteria are needed such as the first or a percentage loss in stiffness. A technique has been developed [16] to continuously monitor the specimen stiffness and the damping properties as indicators of damage from analysis of the waveform of the fatigue load and strain/displacement data at frequent intervals. The data are cross-plotted to obtain hystersis loops of load vs. strain and analysed to obtain the current stiffness and damping properties of the specimen. These data can then be used to determine the failure point based on the required loss in stiffness but there is a greater potential in the use of the data for failure analysis and life prediction.

13.12 DISCUSSION

This Chapter has considered briefly the main test methods and standards relevant to reinforced plastics/composites. The main point to be addressed in all cases is to understand the nature of the material being tested. This aspect is important both in the choice of test method and the interpretation of the results in terms of the specimen response and failure mode. Good progress is being made, and should continue over the next few years, in the production of agreed international standards supported by validation exercises. The precision data obtained will increase the confidence of the designers in the use of these test methods and in the test data produced.

Currently the areas of increased pre-standards activity that will lead to new international test methods are tests for through-thickness properties, in-plane tests for thick sections and physico-chemical test methods. This last group includes DSC (differential scanning calorimetry), DTA (differential thermal analysis), TMA (thermal mechanical analysis), DMA (dynamic mechanical analysis), Chromatographic methods (HPLC, GLC and GPC) and IR (infra-red spectroscopy). These techniques give data on glass transition temperatures, molecular weights etc. for at least one of the stages of in-coming material, B-staged resins or cured laminates. More work is required to establish the correlation of these measurements to the final mechanical performance.

13.13 REFERENCES

1] Sims, G.D., "Development of Standards for Advanced Polymer Matrix Composites" Composites **22** (4), 1991. pp 267-274.

2] Curtis, P.T. "CRAG Test Methods for the Measurement of the Engineering Properties of Fibre Reinforced Plastics", 3rd Edition, RAE Technical Report TR 85099, 1989.

3] Sims, G.D., "Standards for Polymer Matrix Composites, Part II Assessment and Comparison of CRAG Test Methods", NPL Report DMM(M)7, 1990.

4] Sims, G.D., "A VAMAS Round-Robin on Fatigue Test Methods for Polymer Matrix Composites, Part 1 Tensile and Flexural Tests of Unidirectional Material", NPL Report DMM(A)180, 1989.

5] Sims, G.D., "Interim VAMAS Report on Part 1 of Polymer

Composites Fatigue Round-Robin", NPL report DMM(A)69, 1992.

6] Sims, G.D., "Standards for Polymer Matrix Composites, Part I Assessment of CRAG Test Methods Data", NPL Report DMM(M)6, 1990.

7] ISO 5725:1986 (BSI 5497:1987), "Precision of Test Methods, Part 1 Guide for the Determination of Repeatability and Reproducibility for a Standard Test Method by Inter-Laboratory Trial".

8] Carlson, L.A., Pipes, R.B., "Experimental Characterisation of Advanced Composite Materials", Prentice-Hall, New York, 1987.

9] Sims, G.D., "Development of Standards for Advanced Polymer Matrix Composites - a BPF/ACG Overview" NPL Report DMM(A)8, 1990.

10] Hojo, M., Sawada, Y., Miyairi, H., "Effects of Tab Design and Gripping Conditions on Tensile Properties of Unidirectional CFRP in Fibre and Transverse Direction", Proc. Eur. Conf. Comp. Mater., *Composites: Testing and Standardisation*, Amsterdam, 1992.

11] Haeberle, J.G., Matthews, F.L., "A New Test Method for Compression Testing", Proc. Eur. Conf. Comp. Mater., *Composites: Testing and Standardisation*, Amsterdam, 1992.

12] Curtis, P.T., Gates, J., Molyneux, C.G., "An Improved Engineering Test Method for the Measurement of the Compressive Strength of Unidirectional Carbon Fibre Composites", RAE Technical Report TR 91031, 1991.

13] Johnson, A.F., Sims, G.D., "Design Procedures for Plastic Panels" NPL Design Guide, 1987.

14] Sims, G.D., Nimmo, D., Johnson, A.F., Ferriss, D.H., "Analysis of Plate-twist Test for In-plane Shear Modulus", NPL Report DMM(A)54, 1992. (draft test procedure available)

15] Broughton, W.R., Kumosa, M., Hull, D., "An Investigation of Stress Distribution in CFRP Iosipescu Shear Test Specimen", Comp. Sci. Tech., **38**, 1990. pp 299-325.

16] Sims, G.D., Bascombe, D., "Continuous Monitoring of Fatigue Degradation in Composites by Dynamic Mechanical Analysis", Proc. Eur. Conf. Comp. Mater., London, 1988.

The information contained in this chapter was obtained as part of the "Materials Measurement programme", a programme of underpinning research financed by the United Kingdom Department of Trade and Industry. © British Crown Copyright 1993.

APPENDIX: MECHANICAL PROPERTIES OF COMPOSITE MATERIALS' CONSTITUENTS

This appendix contains information on the mechanical properties of the constituents of polymeric composites as well as properties of laminates typically used in a marine context. Information contained in the tables relates to:

1. Moduli, tensile/compressive strengths and strains of thermosetting resins;

2. Moduli and tensile strengths/strains of thermoplastic resins;

3. Moduli, tensile strengths/strains and thermal properties of some reinforcing fibres;

4. Flexural/shear moduli and tensile/compressive/shear strengths of typical laminates;

5. Shear moduli/strengths through-thickness moduli/strengths of typical sandwich core materials;

6. Fire-related properties of metalic and FRP materials.

Table A.1. Typical properties of thermosetting resins.

Material	Specific Gravity	Young's Modulus (GPa)	Poisson's Ratio	Tensile Strength (MPa)	Tensile Failure Strain (%)	Compressive Strength (MPa)	Heat Distortion Temp. (°C)
Polyester (orthophthalic)	1.23	3.2	0.36	65	2	130	65
Polyester (isophthalic)	1.21	3.6	0.36	60	2.5	130	95
Vinyl ester (Derakane 411-45)	1.12	3.4	-	83	5	120	110
Epoxy (DGEBA)	1.2	3.0	0.37	85	5	130	110
Phenolic	1.15	3.0	-	50	2	-	120

Table A.2. Typical properties of some structural thermoplastic resins.

Material	Specific gravity	Young's modulus (GPa)	Tensile yield stress (MPa)	Tensile failure strain (%)	Heat distortion temp. (°C)
ABS (acrylonitrile butadiene styrene)	1.05	3	35	50	100
PET (polyethylene terephthalate)	1.35	2.8	80	80	75
HDPE (high-density polyethylene)	0.95	1.0	30	600–1200	60
PA (polyamide, Nylon 6/6)	1.15	2.2	75	60	75
PC (polycarbonate)	1.2	2.3	60	100	130
PES (polyethersulphone)	1.35	2.8	84	60	203
PEI (polyetherimide)	1.3	3.0	105	60	200
PEEK (polyether-ether ketone)	1.3	3.7	92	50	140

Table A.3. Typical properties of some reinforcing fibres.

Type	Specific Gravity	Young's modulus (axial)* (GPa)	Poisson's ratio*	Tensile strength (GPa)	Failure strain (%)	Coeff. of expansion (axial) (x10-6/°C)	Thermal conductivity (axial) (W/m°C)
E-Glass	2.55	72	0.2	2.4	3.0	5.0	1.05
S2, R-Glass	2.50	88	0.2	3.4	3.5	5.6	-
HS Carbon (Thornel T-40)	1.74	297	-	4.1	1.4	-	-
HS Carbon (Thornel T-700)	1.81	248	-	4.5	1.8	-	-
HS Carbon (Fortafil F-5)	1.80	345	-	3.1	0.9	-0.5	140
HM Carbon (P-75S)	2.00	520	-	2.1	0.4	-1.2	150
HM Carbon (P-12S)	2.18	826	-	2.2	0.3	-	-
Aramid (Kevlar 49)	1.49	124	-	2.8	2.5	-2.0	0.04

* Glass fibres are nearly isotropic; properties of carbon and aramid fibres are strongly anisotropic and are not well defined in the literature.

Table A.4. Typical mechanical properties of FRP laminates.

Material	Fibre volume fraction V_f	Specific gravity	Young's modulus E (GPa)	Shear modulus (GPa)	Tensile strength σ_{UT} (MPa)	Compressive strength (MPa)	Shear strength (MPa)	Specific Young's modulus (E/SG)	Specific tensile strength (σ_{UT}/SG)
E-glass polyester (CSM)	0.18	1.5	8	3.0	100	140	75	5.3	67
E-glass polyester (balanced WR)	0.34	1.7	15	3.5	250	210	100	8.8	147
E-glass polyester (unidirectional)	0.43	1.8	30	3.5	750	600		16.7	417
Carbon/epoxy (high-strength balanced fabric)	0.50	1.5	55	12.0	360	300	110	37	240
Carbon/epoxy (high-strength unidirectional)	0.62	1.6	140	15.0	1500	1300		87	937
Kevlar 49/epoxy (unidirectional)	0.62	1.4	50	8.0	1600	230		36	1143

Table A.5. Typical properties of sandwich core materials.

Core Material	Specific gravity	Shear modulus		Shear Strength		Through-thickness Young's modulus		Through-thickness compressive strength	
		Absolute value (MPa)	Specific value	Absolute value (MPa)	Specific value	Absolute value (MPa)	Specific value	Absolute value (MPa)	Specific value
PVC foam	0.075	25	320	0.8	10.7	50	667	1.1	15
PVC foam	0.13	40	308	1.9	14.6	115	885	3.0	23
PVC foam	0.19	50	260	2.4	12.6	160	842	4.0	21
PU foam	0.10	10	100	0.6	6.0	39	390	1.0	10
PU foam	0.19	30	158	1.4	7.4	83	437	3.0	16
Syntactic foam	0.4	430	1070	-	-	1200	3000	10	25
Syntactic foam	0.8	1000	1250	21	26	2600	3250	45	56
End-grain balsa	0.10	110	1100	1.4	14	800	8000	6	60
End-grain balsa	0.18	300	1670	2.5	14	1400	7780	13	72
Aluminium honeycomb*	0.07	455/205	6500/2930	2.2/1.4	31/20	965	13790	3.5	50
Aluminium honeycomb*	0.13	895/365	6885/2810	4.8/3.0	37/23	2340	18000	9.8	75
GRP honeycomb*	0.08	117/52	1462/650	2.3/1.4	29/18	580	7250	5.7	71
Aramid paper (Nomex) honeycomb	0.065	53/32	815/492	1.7/1.0	26/15	193	2970	3.9	60

* Pairs of numbers refer to longitudinal and transverse directions of hexagonal honeycomb.

Table A.6. Fire-related properties of metallic and FRP materials.

Material	Melting temp. (°C)	Thermal conductivity (W/m°C)	Heat distortion temp. (°C) (BS2782)	Self-ignition temp. (°C)	Flash-ignition temp. (°C)	Oxygen index (%) (ASTM D2863)	Smoke density (DM) (ASTM E662)
Aluminium	660	240	-	-	-	-	-
Steel	1430	50	-	-	-	-	-
E-Glass	840	1.0	-	-	-	-	-
Polyester Resin	-	0.2	70	-	-	20-30	-
Phenolic Resin	-	0.2	120	-	-	35-60	-
GRP (polyester-based)	-	0.4	120	480	370	25-35	750
GRP (phenolic-based)	-	0.4	200	570	530	45-80	75

INDEX

348

Printed in the United States
By Bookmasters